DEVELOPMENTS IN POLYURETHANE—1

THE DEVELOPMENTS SERIES

Developments in many fields of science and technology occur at such a pace that frequently there is a long delay before information about them becomes available and usually it is inconveniently scattered among several journals.

Developments Series books overcome these disadvantages by bringing together within one cover papers dealing with the latest trends and developments in a specific field of study and publishing them within *six months* of their being written.

Many subjects are covered by the series including food science and technology, polymer science, civil and public health engineering, pressure vessels, composite materials, concrete, building science, petroleum technology, geology, etc.

Information on other titles in the series will gladly be sent on application to the publisher.

DEVELOPMENTS IN POLYURETHANE—1

Edited by

J. M. BUIST

F.Inst.P., F.P.R.I.

Director, Anchor Chemical Company Limited, Manchester, UK

APPLIED SCIENCE PUBLISHERS LTD
LONDON

APPLIED SCIENCE PUBLISHERS LTD
RIPPLE ROAD, BARKING, ESSEX, ENGLAND

First edition 1978
Reprinted 1981

British Library Cataloguing in Publication Data

Developments in polyurethane.
1.
1. Polyurethane
I. Buist, Jack Mitchell
620.1'923   TA455.P5

ISBN 0-85334-756-5

WITH 32 TABLES AND 69 ILLUSTRATIONS

© APPLIED SCIENCE PUBLISHERS LTD 1978

Printed in Great Britain by Galliard (Printers) Ltd, Great Yarmouth

FOREWORD

It gave me great pleasure to be invited to write a foreword for this book, since my association with polyurethanes is an active one. I am concerned that the very many developments in this high growth area of the chemical industry should be communicated quickly and widely to all workers in the field.

This book deals with developments over the past 10 years in the major fields in which polyurethanes play an important role. Polyurethane foams and elastomers have captured the imagination and the interest of scientific workers everywhere.

Throughout the world more than 100 000 people are engaged in producing the chemicals for polyurethanes and the finished products. The users of these finished products vastly exceed this number and many people should therefore be interested in the developments described in the present book.

Polyurethane products are produced in a wide variety of forms ranging from flexible through rigid foams to elastomer and cast-solid. Indeed, they are the most versatile combination of chemicals presently used.

Their technology in the USA and Europe has not always followed the same course. In the present book it is interesting to compare the different developments in these continents, particularly in relation to products for the transport industry. Reaction injection moulding (RIM) techniques are described along with the major developments in machinery. There is no doubt that this technique will grow in importance in the many applications for these products in the transport industry in the years ahead.

Rigid foam technology has advanced rapidly in the last 10 years and the main outlet is the construction industry. Urethane rigid foams have the lowest thermal conductivity of any insulating material and all those interested in energy costs and conservation must find these materials exciting. In addition, urethane rigid foams have valuable structural properties. Because of the wide use of both flexible and rigid foams in buildings and in articles in homes, the fire resistance of the finished products is important. A great deal of ill-informed comment has appeared in the press in various countries from time to time and two chapters of this book set out the facts authoritatively. The data given in this book provide an up-to-date picture of a complex subject and show that sufficient information is now available for suitable codes of practice to be issued by government agencies.

The health problems associated with the industrial use of isocyanates is another emotive subject. The polyurethane industry became established in the early 1950s and from its inception the importance of limiting exposure to isocyanates was recognised. The history of the industry using these reactive chemicals in diverse applications is good. The medical effects of exposure to various concentrations of isocyanates are discussed in some detail and reference is made to the guidelines laid down in various countries on the principles to be followed to ensure good industrial hygiene when working with these reactive chemicals.

It gives me great pleasure to recommend this book to all scientific and technological readers interested in these fascinating materials.

A. van Namen
Deputy Chairman
ICI Ltd, Europa Division
Brussels, Belgium

CONTENTS

LIST OF CONTRIBUTORS

W. E. BECKER

Polyurethane Division, Mobay Chemical Corporation, Penn Lincoln Parkway West, Pittsburgh, Pennsylvania 15205, USA.

J. B. BLACKWELL

Polymech Ltd, 2 London Road, Hazel Grove, Stockport SK7 4AH, UK.

F. K. BROCHHAGEN

Bayer AG, PU-Anwendungstechnik, 509 Leverkusen, Bayerwerk, West Germany.

W. BUNGE

Bayer AG, PU-Anwendungstechnik, 509 Leverkusen, Bayerwerk, West Germany.

J. M. BUIST

Anchor Chemical Co. Ltd, Clayton, Manchester M11 4SR, UK.

D. J. DOHERTY

ICI Organics Division, P.O. Box 42, Hexagon House, Blackley, Manchester M9 3DA, UK.

W. GREEN

ICI Organics Division, P.O. Box 42, Hexagon House, Blackley, Manchester M9 3DA, UK.

R. HURD

ICI Organics Division, P.O. Box 42, Hexagon House, Blackley, Manchester M9 3DA, UK.

R. P. REDMAN

ICI Organics Division, P.O. Box 42, Hexagon House, Blackley, Manchester M9 3DA, UK.

R. RUBATTO

Studio Tecnico Mazzucco–Rubatto, Via Vittorio Amemeo 13, 10 121 Turin, Italy.

D. J. WALSH

Department of Chemical Engineering and Chemical Technology, Imperial College of Science and Technology, Prince Consort Road, London SW7 2BY, UK.

G. WOODS

ICI Organics Division, P.O. Box 42, Hexagon House, Blackley, Manchester M9 3DA, UK.

Chapter 1

INTRODUCTION

J. M. BUIST

Anchor Chemical Co. Ltd, Manchester, UK

In 1968 I wrote the following in the Preface to the book *Advances in Polyurethane Technology*:[1]

One of the fascinating features of polyurethanes has been the way that the manufacture and use of these products has been pursued in virtually all the industrial nations of the world. The industry, however, is still a young one and, in general, only those countries which participated in the early inventive work, e.g. Germany, the United Kingdom and the United States of America, have more than a decade of large-scale industrial production experience.

The present book deals with developments over the past ten years, in the major fields where polyurethanes play an important role. The only important application field not covered fully is the surface coating field.

It may be useful to summarise briefly the early history of polyurethanes. In 1945, at the end of the war, B.I.O.S. investigation teams visiting the German chemical industry discovered that in 1937 Dr Otto Bayer (IG Farben Industries) had made an elastomer by reacting tolylene diisocyanate with various polyols. Work had been done to develop both flexible and rigid foams from these elastomers with the main aim of replacing strategic materials or reducing their use, e.g. natural rubber, steel, etc.

During the war ICI had developed a polyester rubber, particularly of interest for impermeability to hydrogen in barrage balloons, using diphenylmethane diisocyanate as the cross-linking agent. In the USA DuPont had also worked extensively with this isocyanate in the elastomer and adhesive fields.

By 1945 when the technology had been developed to meet some war-time applications, e.g. barrage balloons (UK), adding rigidity and strength to aircraft wings with rigid foam (Germany), many of the urgent needs disappeared with the cessation of hostilities. New applications for these versatile materials had to be found.

Those who worked with these materials in the UK and USA were just as surprised as the German workers (when they learned later the full details of the UK and US work) about the similarity of many of the studies made.

In all three countries, because the cross-linking agents used reacted with water to produce carbon dioxide, there were difficulties in producing bubble free films or sheets. Many man-hours were devoted to overcoming this apparent defect. The Germans were the first to devote substantial effort into converting a defect into a virtue. In the late 1940s, Bayer AG worked with Hennecke AG to produce a continuous process for flexible foam and by 1953 they were offering a comprehensive deal covering a production machine, technical know-how and chemicals. The chemicals offered were TDI, a polyester based on adipic acid and diethylene glycol, and a third component containing water, catalyst and surfactants.

At this period ICI, who were producing polyester resins and isocyanates for other purposes, were able to design a machine outside the Bayer patents and in 1955 ICI marketed flexible foam chemicals for use on either the Hennecke high pressure or the ICI low pressure machine.

Developments proceeded quickly and the advent of polyether polyols in America enabled foams to be made from cheaper resins which at the same time had the technical advantages of matching more closely the properties of rubber latex foam and also giving foams of better hydrolytic stability. The polyether polyols soon captured the major part of the flexible urethane foam market for upholstery, seating and bedding but polyester foams retain, even today, a small market share where their particular physical properties meet special requirements.

As has been stated earlier, the first rigid urethane foams were made by IG Farben, during 1940–45, using tolylene diisocyanate. These foams had technical deficiencies and there were toxicity problems associated with this type of system in the processes and applications used at that time. In the early 1950s ICI developed the low toxicity system based on diisocyanate–diphenylmethane compositions[2,3] (MDI) and the first systems using this isocyanate and polyester resins were marketed in 1957. Polyether resins were introduced with this isocyanate in 1959. In the early 1960s claims were made in the USA that crude TDI produced rigid foams as good as those made from MDI, although it should have been clear that crude TDI could never match crude MDI on toxicity grounds. However, rigid foams made from TDI or crude TDI continue to hold a small share of the market today and in the USA crude TDI still has a good share of the refrigerator market. This is one example of how the technologies developed in the USA and in Europe have followed different paths.

Toxicity was important in many of the early experiments with rigid foam because the initial experiments were carried out by crude 'bucket and spade' mixing methods. However, the development of suitable dispensing machines took place quite quickly.[4] The first large-scale applications of the low toxicity MDI rigid foam systems pioneered by ICI were the insulation of portable refrigerated containers, the insulation of the holds and domestic chambers of ships, and the insulation of chemical plant.

The discovery that fluoro-chloro-hydrocarbons could be used in rigid foam recipes as blowing agents, thereby producing foams with thermal insulating properties superior to all known insulating materials, was a major advance.[5] This made possible the first important commercial entry into the appliance market with the thin-walled refrigerator. Because of the widespread production of refrigerators throughout the world, this breakthrough helped to spread the interest in rigid urethane foams to many countries.

In the earlier book[1] it was pointed out that in the early development of urethane foams it was necessary for the chemical suppliers, e.g. Bayer AG and ICI to name two, to develop the special equipment necessary and also to establish the basic principles which should be adopted. Engineering companies, e.g. Hennecke AG (associated with Bayer AG) and Viking Engineering Co. (associated with ICI) produced equipment embodying these principles and also added and developed their own skills.

Many other engineering companies have contributed their expertise and the urethane foam manufacturer now has a wide range of sophisticated machinery of reliable and robust design available for his use. Chapter 9 provides a good summary of the new developments in machinery which have become of interest to industry in the past ten years.

The development of rigid foam laminators described in 1965,[6] was important because new techniques for producing lightweight panels suitable for insulation applications became available. The construction industry in many countries has welcomed the variety of panels which can be produced and many novel methods of using them exist (see Chapter 6).

Manufacturers of rigid foams normally prefer to buy their polyols blended with catalysts, surfactants, etc., as one part of the system and this has led to the establishment of system manufacturers who make fully blended systems to meet the specific needs of their customers. In many cases the system producers have the technical backing of the isocyanate and base polyol manufacturers.

Urethane elastomers were developed early, e.g. Vulkollan,[7] Vulcaprene,[8] Adiprene,[9] and although the early workers all contributed to the advance of knowledge in this field none of these original elastomers established large volume sales. Cast Vulkollan tyres were made in the 1940s and the compression set and other defects should not debar recognition that new techniques of processing these liquid polymers were developed and examined then. Thirty years later, claims have been made in Austria[10] that a suitable cast process to suit the chemicals used has been developed. However, the target sights have been raised in the meantime and cast urethane tyres have a difficult task to match both the performance of radial tyres and the costs of production with low wastage. Nevertheless, urethane pneumatic tyres, as opposed to solid tyres, will one day obtain a share of the tyre market somewhere in the world.

Urethane elastomers have succeeded in establishing worthwhile markets in injection moulded articles (thermoplastic elastomers), shoe soles and in several automobile applications. The use of urethane elastomers in front bumpers, wings and even complete fronts of cars has presented many fascinating technical/commercial problems and the progress made is referred to in Chapter 5. All these developments have encouraged many suppliers to produce a pure form of MDI and in most elastomer formulations either pure MDI or an adduct is used as the cross-linking isocyanate.

Urethane elastomers are also used in many adhesive formulations and in a wide variety of surface coatings. Urethane coatings can be regarded as including isocyanate modified alkyd paints, urethane lacquers and urethane rubbers coated to fabrics or other substrates. In most fields their technical performance has been excellent, but in certain applications their cost has inhibited the development of forecasted market penetrations. However, good technical performance and the establishment of adequate to excellent ageing in service will encourage users to examine periodically the value given by these products.

The *US Foamed Plastics Markets and Directory* for 1977[11] lists eight US producers of TDI with a name plate capacity of 815 million lbs, four US producers of MDI with a name plate capacity of 495 million lbs and eight major polyol producers plus many smaller producers with a named capacity of 1990 million lbs. Two major new MDI producers will come on stream in the USA in 1981 with a combined capacity of 290 million lbs. The statistics given indicate that the US flexible urethane foam market in 1976·

and 1977 is about double that of the US PVC foam market. It is also remarkable that the data published in this report show the market penetration by urethane rigid foams equals the US polystyrene foam market. Production of TDI in Europe (West Germany, UK, France, Italy, Benelux and Spain) appears to be at levels about 15% below that of the US. There are five producers in Japan but the combined Japanese production is probably about 20–25% of the US. Small production units have also operated in the USSR, East Germany and Poland, and plans have been announced for plants to operate in Yugoslavia, Rumania and Bulgaria. TDI is also produced in Mexico and may soon be produced in both Brazil and Argentina. In Europe the MDI capacity and demand are similar to the USA. There are 4 manufacturers with another coming on stream in 1979. In Japan there are four manufacturers of MDI.

The capital investment cost of a TDI (or MDI) plant is high and there is considerable know-how involved in running these plants efficiently and economically. In the case of polyol manufacture the investment cost is much lower and this is one reason why many more manufacturers produce polyols. The major basic raw material is propylene oxide.

The shape of the flexible foam industry has changed with time due to the rapid development of demand for these foams. In most countries, the US and Europe, many of the small manufacturers have either been taken over or gone out of business so that production has tended to be left in the hands of the large powerful international firms. These firms often have flexible foam plants in several countries and in general there has been quick communication of new ideas, process improvements and new market outlets as a result.

The 1977 Directory[11] suggests that rigid urethane foams are expected to grow at 11% per annum up to 1980. The largest market is in building/construction with insulation and energy conservation acting as the motivation. Chapter 6 summarises developments which will contribute and are contributing in this area. Energy conservation studies in factories show that insulation of many items of plant and pipe work is now economical because of the increased cost of energy. The variety of ways in which rigid urethanes can be applied to installed operating plants also helps engineers. Rigid urethane foams are well established in the appliance industry, e.g. refrigerators, deep-freezers, in the industrial countries and the market for foams will grow in line with the growth of these appliances. The increased use of refrigerated transport, trucks, trailers and tank cars will provide a market where greater penetration is still possible.

Rigid urethane foams have also developed useful outlets in decorative furniture components such as drawers, doors, armrests, bed headboards in the US and certain other countries. The use of urethane structural foams in diverse applications such as window frames, TV cabinets and computer housings is increasing. The opportunity exists to extend these developments throughout the world.

Another good market for rigid urethane foams is marine insulation ranging from large tankers and conventional ships, to small boats. Polystyrene foam blocks are still widely used as flotation material in small boats but urethane foam has the advantage of giving extra structural rigidity to the hull. An interesting development is the use of urethane foam in marine salvage to float wrecks. Data becoming available show worthwhile savings in time and cost over more conventional salvage methods.[11]

There is no doubt that many outlets for rigid urethane foams have still to be discovered in other industries. However, as indicated above, rigid urethane foams have become accepted by several established industries. The experience gained in the production techniques and methods of fabrication, in meeting the very specific needs of these industries, is available to extend their use in many other fields. Work being done on developing structural foams based on urethanes will bear fruit in the future.

Because construction and building was a major outlet for urethanes it has been necessary for the urethane industry to establish close contact with architects and builders. These close contacts must continue in order to improve the knowledge of urethane foams potential, properties, methods of application, etc. so that architects and builders can plan to use the material intelligently and safely. From time to time there have been scares about the fire risks involved with urethane foams and indeed with all polymeric materials. The data accumulated by the urethane industry to establish the true facts about the risks involved and refute the irresponsible statements sometimes made are impressive. The data on rigid foams are dealt with in Chapter 8 and Chapter 4 covers flexible foam. These chapters should put this whole emotive subject into proper perspective and are important contributions to this book.

Another topic of major importance is that of health and safety and Chapter 10 is an up-to-date summary of present knowledge. The industry has a good record in spite of dealing with very reactive chemicals. The procedures for handling the chemicals safely are laid down clearly and specifically. Because the high reactivity of the chemicals was known from the start, much more work has been carried out to study the medical effects of the chemicals on men and women working with them and this extensive work provides the basis for the reassurances given in Chapter 10. Reference is also made to fairly extensive animal pathology work reported in the literature but it is as well to remember that it is not always valid to extrapolate from animal results. Animal effects are not always produced in humans when exposed to the same concentrations.

The reader of the book will also find much to interest him in other chapters such as the growing application of microcellular urethanes (Chapters 3 and 4) and in Chapter 9 on machinery.

More than 100 000 people are working with urethane chemicals throughout the world and these developments have been accomplished in a brief spell of 25 years of commercial development. The prospect in the years ahead is just as exciting.

REFERENCES

1. BUIST, J. M. and GUDGEON, H. (1968). *Advances in Polyurethane Technology*, Maclaren & Sons Ltd, London.
2. IMPERIAL CHEMICAL INDUSTRIES LTD, British Patent 848,671.
3. BUIST, J. M. and HURD, R. (1959). *Proc. Inst. Refrig.*, 60.
4. BUIST, J. M. (1959). *Proc. International Rubber Conference*, Washington.
5. GENERAL TIRE & RUBBER CO., British Patent 821,342.
6. BUIST, J. M. (1965). *J. Cell. Plast.*, 1(1), 101.
7. IG FARBENINDUSTRIE AG, German Patent 728,981.
8. HARPER, D. A., NAUNTON, W. J. S., REYNOLDS, R. J. W. and WALKER, E. E. (1947). Proc. XIth Cong. Pure and App. Chem. London 5, 311.
9. E. J. DUPONT DE NEMOURS AND CO., British Patent 533,733.
10. *European Rubber Journal*, Apr. 1976, 10.
11. *US Foamed Plastics Markets and Directory* (1977). Technomic Publishing Co. Inc.

Chapter 2

NEWER SYNTHETIC ROUTES TO ISOCYANATES AND URETHANES

D. J. WALSH

Department of Chemical Engineering and Chemical Technology, Imperial College of Science and Technology, London, UK

SUMMARY

This chapter describes several new routes for the preparation of isocyanates, in situ isocyanate generators and polyurethanes which have been developed over the last 10 or 15 years. Some of these routes have found commercial applications although many are still at the research stage. The chapter concludes by describing several new polymers which may be made from isocyanates.

INTRODUCTION

The standard procedures for making these compounds have been fully described elsewhere[1] and it is therefore necessary to do no more than briefly review them here.

Polyurethanes can be prepared by the reaction of a di- or polyisocyanate with a di- or polyfunctional alcohol, e.g.

$$n(OCN—R—NCO) + n(HO—R'—OH) \longrightarrow$$

$$OCN(\!\!-\!\!R—NH—CO—O—R'—O—CO—NH\!\!-\!\!)_{n-1}R$$

$$HO—R'—O—CO—NH$$

The most important isocyanates used are tolylene diisocyanate (TDI), the commercial material usually being an 80:20 mixture of the 2,4 (I) and

2,6 (II) isomers, and polymeric MDI (III).[2] The latter is being used to an increasing extent due to the fact that it is less volatile than TDI.

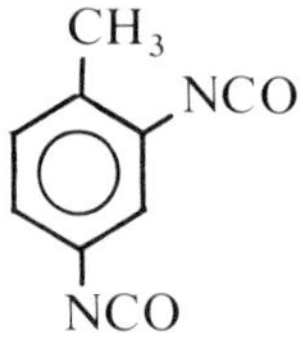

(I)

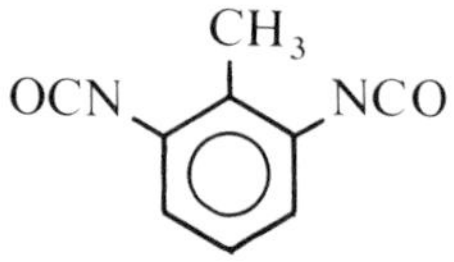

(II)

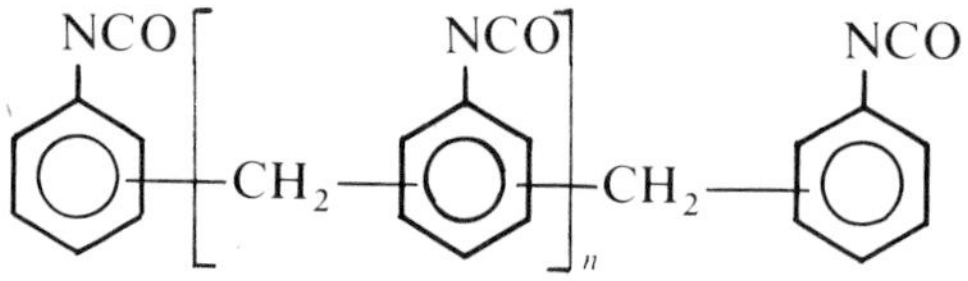

(III) $n = 0, 1, 2, 3$

Other commonly used isocyanates include 4,4′diphenylmethane diisocyanate (MDI) (IV) which is also a major component of polymeric MDI ($n = 0$), and the aliphatic diisocyanate, hexamethylene diisocyanate (HMDI) (V)

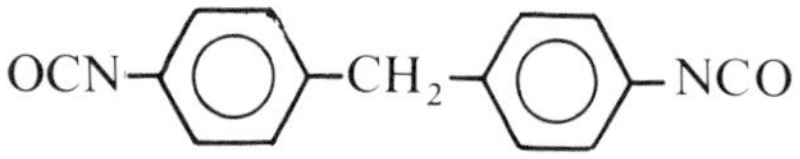

(IV)

$$OCN{-}CH_2{-}CH_2{-}CH_2{-}CH_2{-}CH_2{-}CH_2{-}NCO$$

(V)

The alcohols commonly used include hydroxyl ended polypropylene oxides (VI) as well as hydroxyl ended polyesters and polytetramethylene oxides, and short chain alcohols such as butane-1, 4-diol (VII) and trimethylolpropane (VIII)

$$HO\left(\underset{\displaystyle CH_3}{\overset{|}{CH}}{-}CH_2{-}O\right)_n CH_2{-}CH_2\left(O{-}CH_2{-}\underset{\displaystyle CH_3}{\overset{|}{CH}}\right)_{n'} OH$$

(VI)

$$HO{-}CH_2{-}CH_2{-}CH_2{-}CH_2{-}OH$$

(VII)

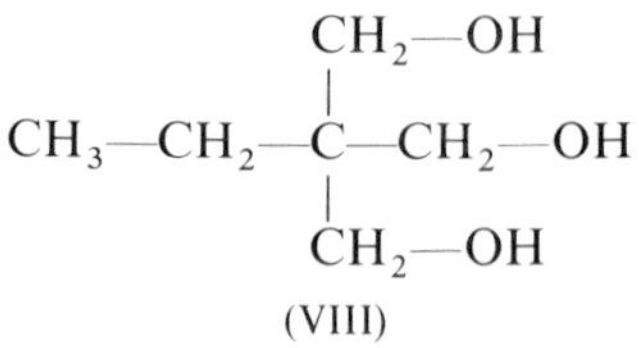

(VIII)

The polymers can be made into three dimensional networks by the inclusion of polyisocyanates or triols in the mixture, or by using diisocyanate in excess of the stoichiometrically required amount, when further reactions such as allophanate formation can occur by reaction of excess isocyanates with already formed urethane groups at higher temperatures

$$\begin{array}{ccc} & & \overset{\displaystyle |}{NH} \\ \overset{\displaystyle |}{NCO} & & \overset{\displaystyle |}{CO} \\ & \rightarrow & \overset{\displaystyle |}{} \\ -O-CO-NH- & & -O-CO-N- \end{array}$$

Polyurethane formation is catalysed by many compounds including tertiary amines and organotin compounds.

Linear polyurethane elastomers containing flexible polyols and low molecular weight diols exhibit phase separation consistent with the behaviour of block copolymers, and the physical cross-links thus formed, being thermally reversible, allow a mouldable polymer to be made.

Such formulations intended to form a permanent polymer network have to be supplied as a two-component system and several ways of supplying a one-component system have been tried. Blocked isocyanates have been used which only become active at elevated temperatures, such as the phenol adducts (IX) (phenol urethanes). The fact that a volatile by-product is formed has restricted the use to coating applications, and in this case the phenol produced would be a considerable hazard

Some of the research into new routes to polyurethanes has been directed at systems which would overcome this problem and allow one-component systems to produce polyurethanes at elevated temperatures without any unwanted by-product being formed.

Isocyanates can be prepared by Curtius, Hofmann and Lossen rearrangements.[4]

$$R.CO.Cl \xrightarrow{NaN_3} R.CO.N_3 \xrightarrow{-N_2} R.NCO$$

$$R.CO.NH_2 \xrightarrow{NaOBr} R.CO.NH.Br \xrightarrow{-HBr} R.NCO$$

$$R.CO.OCH_3 \xrightarrow[-CH_3OH]{NH_2OH} R.CO.NH.OH \xrightarrow{-H_2O} R.NCO$$

These reactions are not suitable for large-scale production. They may also be prepared by the cyanation process where, for example, an alkyl halide is treated with a metal cyanate. Particular interest has been shown in the reactions of chloromethylated aromatics,[3] e.g.

Such reactions may find uses for special applications.

Isocyanates are more generally prepared by phosgenation of the respective amines

$$R—NH_2 + COCl_2$$

$$\Downarrow$$

$$R—NCO + 2HCl$$

The phosgene itself is made from carbon monoxide and chlorine. Any process which uses chlorine and produces hydrochloric acid as a by-product uses expensive materials and, unless the hydrochloric acid can be used in other processes, is inefficient. The tendency throughout industry over the last 10 or 20 years has been to replace such processes; for example, in phenol and alkylene oxide productions. This provides the impetus for some alternative synthetic routes to isocyanates.

Tolylene diisocyanate is prepared by nitration of toluene to dinitrotoluene, reduction to tolylene diamine, and phosgenation to the diisocyanate.

Polymeric MDI is prepared by the phosgenation of polymethylene polyphenyl amine in chlorobenzene solvent. The amine is an aniline–formaldehyde condensation product (X) prepared via aniline hydrochloride.[2]

$$(n + 2) \quad \text{NH}_3{}^+\text{Cl}^- \text{ ring} \quad + (n + 1)\,\text{HCHO}$$

(X)

The mechanisms of production of polymeric MDI and TDI have recently been described and reviewed in detail.[4]

Aliphatic isocyanates have an advantage over aromatic isocyanates in that the polyurethanes produced from them do not yellow on exposure to air and light. This makes them better for certain applications. HMDI, however, has a serious disadvantage in that it has a high vapour pressure, making it one of the most hazardous of diisocyanates. Non-volatile aliphatic diisocyanates have been developed; the main problem being to obtain an inexpensive aliphatic diamine.

Xylylene diisocyanates have been prepared by ammoxidation of xylenes followed by hydrogenation and phosgenation[5]

Other diisocyanates have been prepared based on isophorone,[6] which can be prepared by the condensation of acetone.

The synthesis of many diisocyanates by the phosgenation route has been reviewed in detail.[7]

Any alternative route to isocyanates or polyurethanes which would be economically viable, would have to compete with the two major described routes or offer some other advantage over the existing routes, such as in the processing conditions or the physical properties of the product.

In the following sections several different new routes to isocyanates and polyurethanes are described. No particular significance should, however, be given to the order of presentation.

CARBONYLATION

In the search to find cheaper routes to isocyanates it was found that aryl isocyanates could be formed by the direct reaction of carbon monoxide with aromatic nitro compounds. It was first reported, in 1967, that phenyl isocyanate could be prepared from nitrobenzene by reaction with carbon monoxide at 190 °C and 500 atmospheres pressure in the presence of rhodium on carbon, and ferric chloride or other Lewis acids. The yield was only of the order of 35%.[8]

The production of tolylene diisocyanate from dinitrotoluene has been reported in 57% yield[9] and later 76% yield[10] and 80·6% yield[11] using catalysts of palladium, iron and molybdenum compounds.

It seems possible that a satisfactory commercial process for the production of tolylene diisocyanate by this method will be developed. The route removes one step from the process and the obvious reduction in raw material and capital costs will make such a route potentially much more economical. At present, this method requires large amounts of expensive catalysts which are difficult to recover.

FUROXANS

Furoxans (furazan N-oxides) are compounds containing the ring structure (XI) which can be formed by the dimerisation of nitrile oxides. When some of these are heated they decompose to give isocyanates.[12] The route is believed to be via the nitrile oxide followed by rearrangement to the isocyanate.[13]

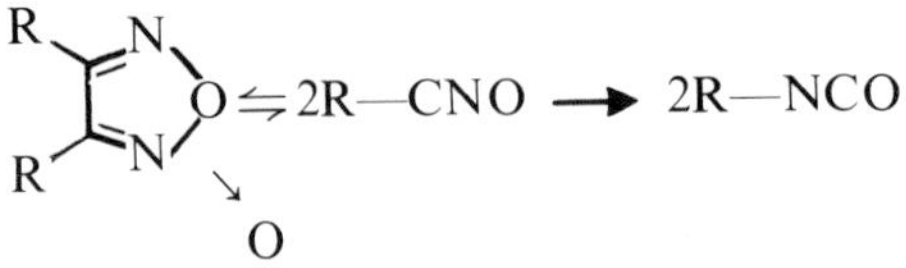

$$R-C \underset{R'}{\overset{N}{\diagdown}} \overset{N}{\diagup} O \rightleftharpoons 2R-CNO \longrightarrow 2R-NCO$$

(XI)

Thus, when the diphenyl compound (XI; R = Ph) was heated, phenyl isocyanate distilled off in a yield of 42 %. Alternatively, when it was heated under reflux in dodecan-1-ol at 257 °C the carbamate (XII) was formed in 81 % yield. That the route proceeds via a nitrile oxide was suggested by the fact that when the diphenyl compound was heated at 245 °C in tetradec-1-ene, the isoxazoline (XIII) was isolated.

$$C_{12}H_{25}-O-CO-NH-Ph$$

(XII)

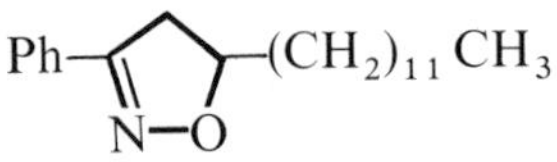

$$Ph-C \underset{N-O}{\diagdown} (CH_2)_{11}CH_3$$

(XIII)

Furoxans, when required as a source of nitrile oxides for isocyanate formation, may be prepared by the reaction of N_2O_3 with olefins to form a pseudo nitrosite (XIV).[14] The pseudo nitrosite, on heating, rearranges to the 2-nitro oxime (XV), which on dehydration gives a furoxan. Alternatively, the furoxan may be prepared directly from the pseudo nitrosite.[14]

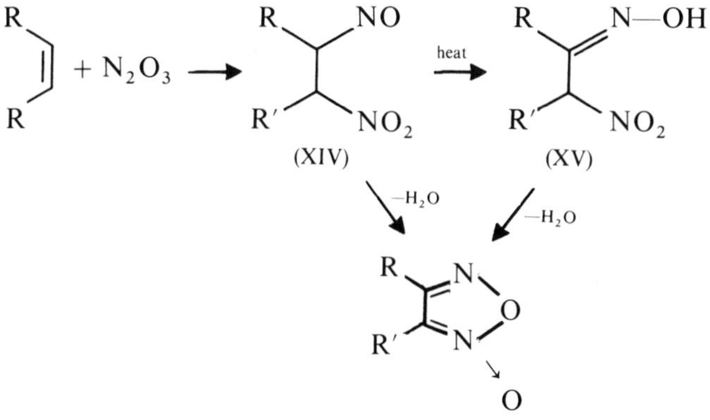

$$R\!\!\diagdown\!\!\diagdown \quad + N_2O_3 \longrightarrow \quad \underset{R'}{R}\!\!\diagdown\!\!\diagup^{NO}_{NO_2} \quad \overset{heat}{\longrightarrow} \quad \underset{R'}{R}\!\!\diagdown\!\!\diagup^{N-OH}_{NO_2}$$

(XIV)   (XV)

$$-H_2O \diagdown \qquad \diagup -H_2O$$

$$\underset{R'}{R}\!\!\diagdown\!\!\diagup\!\!\diagdown\!\!\overset{N}{\underset{N}{\diagdown}}\!\!\overset{O}{\diagup}$$

O

Thus, when butene was dissolved in a mixture of ether and light petroleum and a mixture of nitric oxide and air passed through at $-10\,°C$, the pseudo nitrosite (3-nitro-2-nitrosobutane) was formed in 29 % yield. On heating in dimethyl sulphoxide for 10 min at 115 °C this rearranged to give 3-nitro-butan-2-one oxime. When this was heated at 110 °C for 15 min in polyphosphoric acid, 3,4-dimethylfuroxan (XVI) was formed.

These reactions can also be performed using cyclic olefins such as cyclohexene, in which case the product would be 3,4-tetramethylene-furoxan (XVII).

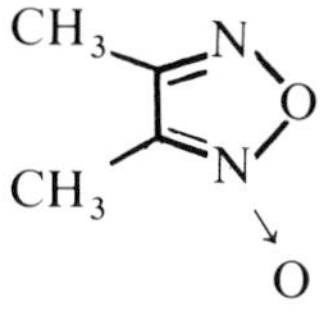

(XVI) 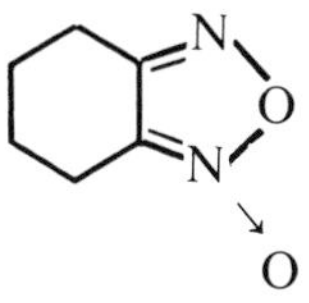   (XVII)

When furoxans of type (XI) are thermally decomposed they produce monoisocyanates, whereas it was found that the bicyclic furoxans such as (XVII) will produce diisocyanates,[15] and when heated with alcohols they produce biscarbamates.

Thus, when (XVIII) is heated in decan-1-ol at 228 °C, the biscarbamate (XIX) can be isolated in 70 % yield, whereas thermolysis of (XVIII) yields 1,10-diisocyanatodecane (XX).

In a similar way 2,4-diisocyanatobicyclo (3.3.0) oct-6-ene was prepared from (XXI) by refluxing in dry toluene saturated with SO_2 in 78 % yield.[16]

(XVIII)   (XXI)

$$CH_3(CH_2)_9\!-\!O\!-\!CO\!-\!NH\!-\!(CH_2)_{10}\!-\!NH\!-\!CO\!-\!O\!-\!(CH_2)_9CH_3$$
(XIX)

$$O\!=\!C\!=\!N\!-\!(CH_2)_{10}\!-\!N\!=\!C\!=\!O$$
(XX)

The potential benefits of furoxans as a route to polyurethanes are that they allow (a) a series of diisocyanates to be prepared which are not easily prepared by other routes, and (b) urethanes to be prepared at high temperatures without the evolution of by-products, as in the case of blocked isocyanates.[17] The high temperature necessary for this to occur could, however, be a problem.

ISOCYANATES FROM UREAS

The reaction of phosgene with 1,3-disubstituted ureas has been known for some time. Recently this reaction has been used to synthesise isocyanates which cannot be prepared directly by the phosgenation of an amine. These and other related reactions of ureas have been comprehensively reviewed and described.[22] The amine is first converted to the urea by reaction with a readily obtainable isocyanate, and the urea is then treated with phosgene to produce two molecules of isocyanate.

$$R\!-\!NH_2 + R'\!-\!NCO \rightarrow R\!-\!NH\!-\!CO\!-\!NH\!-\!R'$$

$$R\!-\!NH\!-\!CO\!-\!NH\!-\!R' + COCl_2 \rightarrow R\!-\!NCO + R'\!-\!NCO + 2HCl$$

Thus toluene-p-sulphonamide can be converted to the sulphonyl-isocyanate by phosgenation of 1-p-toluenesulphonyl-3-n-butylurea at a temperature of 80–100 °C. It is also possible to phosgenate toluene-p-sulphonamide directly with phosgene using n-butylisocyanate as a catalyst.[19]

1,2-Alkylenediisocyanates can be prepared by the reaction of a cyclic urea with phosgene.[20] The cyclic ureas (2-imidazolidinones) can be prepared from the reaction of a 1,2-alkylenediamine with carbon dioxide. The diamine cannot be directly reacted with phosgene as this leads to polymeric products.[21]

Thus ethylene diisocyanate (XXII) can be prepared from 2-imidazolidinone (XXIII) by reaction with phosgene in 1,2-dichloroethane at 75–80 °C to form 2-imidazolidinone-1-carbonyl chloride (XXIV). This is treated with triethylamine at room temperature to produce the diisocyanate.

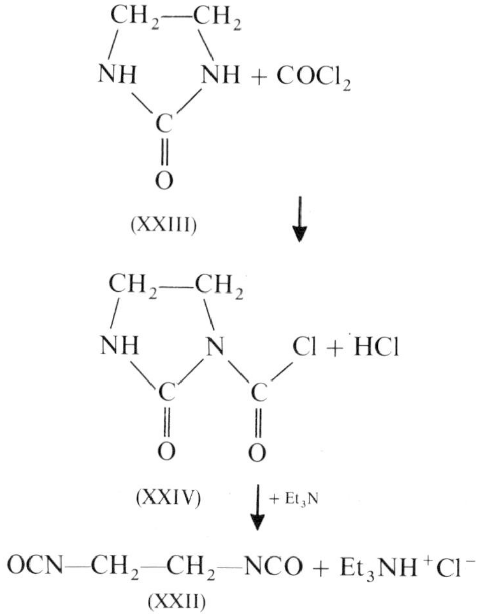

$$CH_2\!\!-\!\!CH_2$$

NH   NH + COCl$_2$

C

O

(XXIII)

$$CH_2\!\!-\!\!CH_2$$

NH   N   Cl + HCl

C   C

O   O

(XXIV)   + Et$_3$N

$$OCN\!-\!CH_2\!-\!CH_2\!-\!NCO + Et_3NH^+Cl^-$$
(XXII)

Long-chain alkylenediisocyanates can be prepared by the phosgenation[19] of the corresponding alkylenebis-ureas. Such diisocyanates are often difficult to prepare because of the low solubility of the amine hydrochlorides. A long reaction time and a large excess of phosgene is required and this leads to unwanted side reactions taking place.[21] A route via the urea overcomes this problem.

Thus, if hexamethylene diamine is treated with toluene-*p*-sulphonylisocyanate a bis-urea (XXV) is formed which can be treated with phosgene at 130–140 °C to give hexamethylene diisocyanate in 95·5 % yield. Toluene-*p*-sulphonylisocyanate can be recovered in 67% yield.

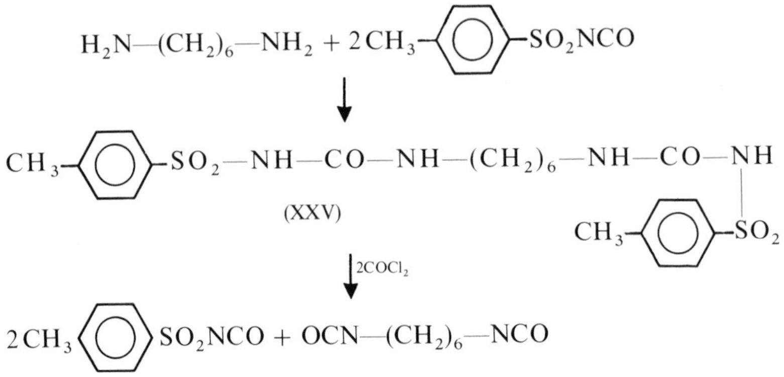

$$H_2N-(CH_2)_6-NH_2 + 2CH_3-\underset{}{\bigcirc}-SO_2NCO$$

$$\downarrow$$

$$CH_3-\underset{}{\bigcirc}-SO_2-NH-CO-NH-(CH_2)_6-NH-CO-NH$$

(XXV)

$$CH_3-\underset{}{\bigcirc}-SO_2$$

$$\downarrow 2COCl_2$$

$$2CH_3\underset{}{\bigcirc}SO_2NCO + OCN-(CH_2)_6-NCO$$

Toluene-2,4 disulphonyldiisocyanate can also be prepared from the bis(*n*-butylurea).[19]

m-Phenylene-disulphonyldiisocyanate has also recently been prepared by phosgenation of the disulphonamide using butylisocyanate as a catalyst.[18]

This general method for the preparation of diisocyanates would be unlikely to find any general application because it involves an extra step in the synthesis. It may however find application in special and specific circumstances.

AMINIMIDES

Aminimides such as trimethylamine benzimide (XXVI) decompose on heating to give an isocyanate and a tertiary amine. The amine is a catalyst for further reactions of the isocyanate and in the case of aromatic isocyanates promotes the formation of trimer.[23]

$$C_6H_5-CO-N^--N^+(CH_3)_3 \longrightarrow C_6H_5-NCO + (CH_3)_3N$$

(XXVI)

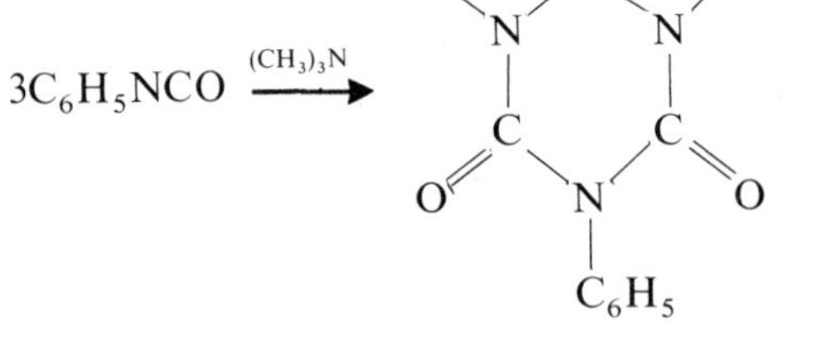

$$3C_6H_5NCO \xrightarrow{(CH_3)_3N}$$

Aminimides can be prepared by the reaction of an acid chloride with 1,1-dimethylhydrazine followed by quaternisation and neutralisation of the salt.[24]

$$R\,COCl + H_2N\,N(CH_3)_2 \rightarrow R\,CONHN(CH_3)_2 + HCl$$

$$R\,CONHN(CH_3)_2 + CH_3Cl \rightarrow R\,CONH\overset{+}{N}(CH_3)_3\overset{-}{Cl}$$

$$R\,CONH\overset{+}{N}(CH_3)_3\overset{-}{Cl} + NaOH \rightarrow R\,CO\overset{-}{N}\!-\!\overset{+}{N}(CH_3)_3 + NaCl + H_2O$$

Alternatively, they can be prepared by the reaction of an ester with 1,1,1-trimethylhydrazinium salts in the presence of anhydrous bases. This route is believed to be via 1,1,1-trimethylaminimine.[23]

$$H_2N\,\overset{+}{N}(CH_3)_3\,\overset{-}{Cl} + CH_3ONa \rightarrow H\,\overset{-}{N}\,\overset{+}{N}(CH_3)_3 + NaCl + CH_3OH$$

$$R\,CO_2CH_3 + H\,\overset{-}{N}\,\overset{+}{N}(CH_3)_3 \rightarrow R\,CO\,\overset{-}{N}\,\overset{+}{N}(CH_3)_3 + CH_3OH$$

They can also be generated by the reaction of 1,1-dimethylhydrazine with an epoxide and an ester,[25-27] thus producing a potentially self-reacting molecule.

$$(CH_3)_2\,N\,NH_2 + R\!-\!\overset{\diagdown}{CH}\!-\!\overset{\diagup}{CH_2} \longrightarrow R\,CHCH_2\!-\!\overset{+}{N}(CH_3)_2$$

(with O in epoxide ring, and substituents $\overset{-}{O}$ above, NH_2 below on product)

$$R'CO\,N\,\overset{-}{N}\!-\!CH_2\,CHOH\,R \xleftarrow{R'COOCH_3} R\,CH\!-\!CH_2\!-\!\overset{+}{N}\!-\!\overset{-}{NH}$$

(with CH$_3$ groups on nitrogens and OH group)

Bisaminimides can be prepared using difunctional precursors, and these have been used as isocyanate generators. Thus, using isophthaloyl chloride and the first of these synthetic routes, bis(trimethylamine) isophthalimide (XXVII) is obtained. When this compound was heated at 240 °C with a hydroxyl ended polyester in the presence of stannous octoate, trimethylamine was evolved and a polymeric elastomer was formed which showed an infra-red spectrum characteristic of a polyurethane.[23]

Bis(NN-dimethyl-2-hydroxypropylamine)adipimide, and bis(NN-dimethyl-2-hydroxypropylamine) sebacimide are used as tyre-cord adhesives.[29]

$$(CH_3)_3N^+ \!-\! N^- \!-\! OC\!-\!\underset{\text{(XXVII)}}{\bigodot}\!-\!CO\!-\!N^- \!-\! N^+(CH_3)_3$$

Much interest has been shown in the uses of aminimides containing polymerisable vinyl groups such as 1,1-dimethyl-1-(2-hydroxypropyl)-amine methacrylimide (XXVIII).

$$CH_3\!-\!CHOH\!-\!CH_2\!-\!\overset{\overset{\displaystyle CH_3}{|}}{\underset{\underset{\displaystyle CH_3}{|}}{N^+}}\!-\!\overset{-}{N}\!-\!CO\!-\!\overset{\overset{\displaystyle CH_3}{|}}{C}\!=\!CH_2$$

(XXVIII)

It is possible to copolymerise such compounds with a wide range of vinyl monomers. Copolymers of (XXVIII) and styrene mixed with bishydroxy compounds form urethane cross-linked polymers[28] on heating. A recent review describes the many applications which such compounds have found in the fields of tyre-cord adhesives, surface coatings and textile coatings.[29]

ISOCYANATES VIA HYDROXAMIC ACID HALIDES

A route has been reported for the preparation of aromatic isocyanates from aromatic compounds containing one or more methyl or halomethyl groups.[30] The compound is treated with a nitrosyl halide using visible or UV irradiation as the initiator. A hydroxamic acid halide is generated which can be thermally decomposed to give the isocyanate.

Nitrosyl chloride can be formed *in situ* from mixtures of nitric oxide and chlorine. It acts as a chlorinating agent and thus both methyl and chloromethyl groups are converted to hydroxamoyl chlorides.

$$Ph\!-\!CH_3 \xrightarrow[h\nu]{NOCl} Ph\!-\!CH\!=\!NOH$$

$$\downarrow Cl_2$$

$$Ph\!-\!CH_2Cl \xrightarrow[h\nu]{NOCl} Ph\!-\!CCl\!=\!NOH$$

$$-HCl \downarrow 180°C$$

$$Ph\!-\!NCO$$

ISOCYANATES VIA MONOTHIOCARBAMATES

A recent synthesis of isocyanates has been described.[31] An aliphatic amine is treated with carbonyl sulphide and triethylamine to form the monothiocarbamate salt (XXXI). The salt is condensed with S-ethyl chlorothioformate and the product (XXXII) thermally decomposed to yield the isocyanate.

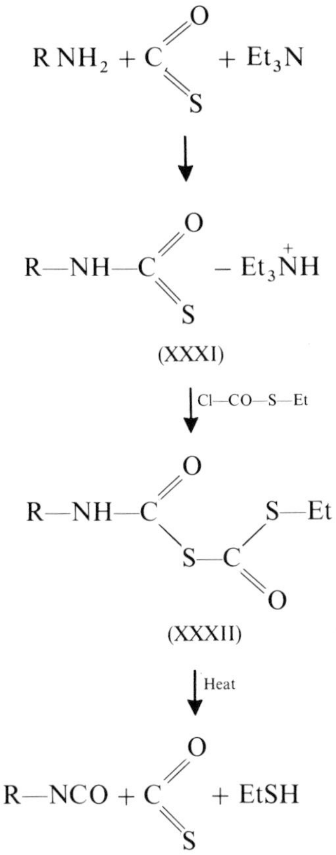

$$R\,NH_2 + C\overset{O}{\underset{S}{\Bigg\langle}} + Et_3N$$

$$\downarrow$$

$$R{-}NH{-}C\overset{O}{\underset{S}{\Bigg\langle}} - Et_3\overset{+}{N}H$$

(XXXI)

$$\Big\downarrow \text{Cl—CO—S—Et}$$

$$R{-}NH{-}C\overset{O}{\underset{S{-}C}{\Big\langle}}\underset{O}{\overset{S{-}Et}{\Big\rangle}}$$

(XXXII)

$$\Big\downarrow \text{Heat}$$

$$R{-}NCO + C\overset{O}{\underset{S}{\Bigg\langle}} + EtSH$$

Thus phenethylisocyanate was prepared in 37–59% yields and hexamethylene diisocyanate in 52% yield. This method is unlikely to be favoured except in very special circumstances.

If S-ethyl chlorothioformate is considered as a condensation product of phosgene with ethanethiol then the overall reaction is equivalent to the reaction of phosgene with an amine.

SUBSTITUTED 1,3,2,4-DIOXATHIAZOLE S-OXIDES AND RELATED COMPOUNDS

Isocyanates have been prepared from 1,3,2,4-dioxathiazole S-oxides (XXXIII). These can be prepared by the reaction of thionyl chloride with hydroxamic acids.[32]

$$\text{R—CO—NH—OH} + \text{SOCl}_2 \rightarrow \underset{(\text{XXXIII})}{\text{R—C=N}} + 2\,\text{HCl}$$

When heated they decompose to give sulphur dioxide and an isocyanate.

$$\underset{(\text{XXXIII})}{\text{R—C=N}} \rightarrow \text{R—NCO} + \text{SO}_2$$

R can be aliphatic or aromatic.

Other related compounds can be prepared from the reactions of phosgene and oxalyl chloride with the hydroxamic acids.[33]

$$\text{R CO=NH—OH} + \text{COCl}_2 \rightarrow \underset{(\text{XXXIV})}{\text{R—C=N}}$$

$$\text{R CO—NH—OH} + \text{Cl CO CO Cl} \rightarrow \text{R—C=N}$$

Such compounds also decompose to give the isocyanate and carbon dioxide or carbon dioxide plus carbon monoxide, respectively. The first of these is favoured in practical circumstances because the carbon dioxide produced is non-poisonous. The compound (XXXIV) is commonly referred to as a nitrile carbonate.

When polyfunctional compounds such as adipo di(nitrile carbonate) are heated at 100 °C with polyols (such as polytetramethylene ether glycol) with suitable catalysts, a high molecular weight polyurethane is formed.[34]

This route does not represent a cheaper one to polyurethanes but shows potential as an isocyanate-generating system which cures at reasonable temperatures. The carbon dioxide generated can also be used to produce a polyurethane foam.

ROUTES TO POLYURETHANES NOT USING ISOCYANATES

Apart from the reaction of an isocyanate with an alcohol, it has long been known that polyurethanes can also be formed by the reaction of a diamine with a bischloroformate. This reaction has not found much practical use.

$$n\,Cl.CO.OR\,O.CO.Cl + n\,H_2N\text{---}R'\text{---}NH_2$$
$$\downarrow$$
$$\text{---}(NH.CO.O.R\text{---}OCONH\text{---}R'\text{---})_{\overline{n}} + 2n\,HCl$$

Recently there has been interest in the copolymerisation of aziridines and carbon dioxide to form copolymers of polyurethanes and polyaziridines[35]

$$(m+n)\,R\text{---}\underset{\underset{R_1}{|}}{C}\underset{}{\overset{\overset{\displaystyle R_4}{\underset{\displaystyle N}{|}}}{\diagup\!\diagdown}}\underset{\underset{R_3}{|}}{C}\text{---}R_2 + m\,CO_2$$

$$\downarrow$$

$$\left(\!\!\begin{array}{c} R_1\ R_2\ R_4\ O \\ |\ \ |\ \ |\ \ || \\ \text{---}C\text{---}C\text{---}N\text{---}C\text{---}O\text{---} \\ |\ \ | \\ R\ \ R_3 \end{array}\!\!\right)_{m}\!\!\left(\!\!\begin{array}{c} R_1\ R_2\ R_4 \\ |\ \ |\ \ | \\ \text{---}C\text{---}C\text{---}N\text{---} \\ |\ \ | \\ R\ \ R_3 \end{array}\!\!\right)_{n}$$

A wide variety of aziridines have been used in this way. It was originally thought that ethyleneimime itself would not copolymerise but this was later shown not to be the case.[36] The copolymers are reported to contain short blocks of each kind of unit.

The polymerisation is carried out, often in the presence of catalysts, at high carbon dioxide pressures. A higher excess of carbon dioxide results in a higher concentration of urethane groups in the product.

OTHER POLYMERS FROM ISOCYANATES

So far in this chapter isocyanates have only been discussed in terms of their usefulness for polyurethane production, but many other types of polymers can be made from isocyanates.

The trimerisation of isocyanates has been utilised to form isocyanurate polymers, and the polymerisation of isocyanates to form a polyamide or 1-nylon (XXXV) has been known for some time.

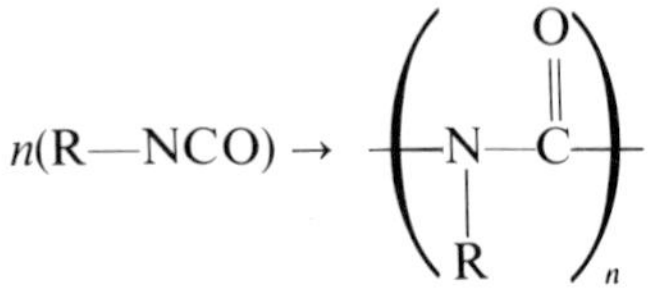

$$n(\text{R}-\text{NCO}) \rightarrow \left(\begin{array}{c} \text{O} \\ \| \\ -\text{N}-\text{C}- \\ | \\ \text{R} \end{array} \right)_n$$

(XXXV)

Diisocyanates can also react to form polycarbodiimides (XXXVI).[38]

$$n\,\text{OCN}-\text{R}-\text{NCO} \xrightarrow{\text{catalyst}} (\text{R}-\text{N}=\text{C}=\text{N})_n + n\,\text{CO}_2$$

(XXXVI)

and the reaction of diisocyanates with diepoxides produces polyoxazolidinones (XXXVII).[39]

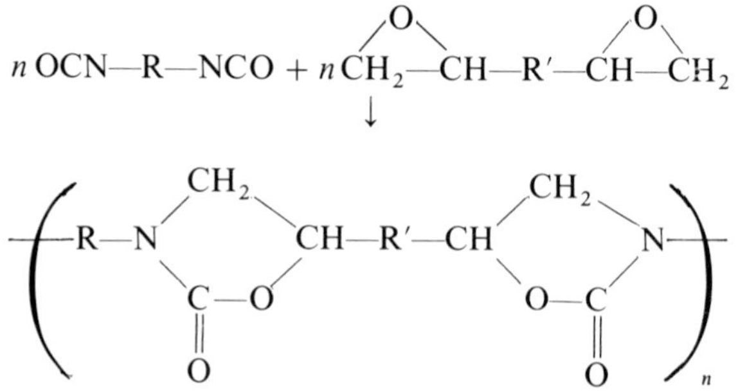

$$n\,\text{OCN}-\text{R}-\text{NCO} + n\,\underset{\text{CH}_2-\text{CH}}{\overset{\text{O}}{\diagup\diagdown}}-\text{R}'-\underset{\text{CH}-\text{CH}_2}{\overset{\text{O}}{\diagup\diagdown}}$$

$$\downarrow$$

$$\left(-\text{R}-\text{N} \underset{\text{C}-\text{O}}{\overset{\text{CH}_2}{\diagup\diagdown}} \text{CH}-\text{R}'-\text{CH} \underset{\text{O}-\text{C}}{\overset{\text{CH}_2}{\diagup\diagdown}} \text{N}- \right)_n$$

(XXXVII)

Several new types of polymers have recently been prepared and reviewed.[37,40] Many of these polymers show very good temperature-resistant properties.

One recently introduced class of polymers are the poly(parabanic acids).[41] When isocyanates condense with hydrogen cyanide, imidazol-idines (XXXVIII) are formed, which yield the polyparabanic acid (XXXIX) on hydrolysis.

$$n \, OCN—R—NCO + n \, HCN \rightarrow$$

(XXXVIII)   (XXXIX)

The polymer based on MDI gives a glass transition temperature of 289 °C. Tough films can be prepared from solutions of these polymers.

The reaction of isocyanates with acid anhydrides to produce polyimides has been known for some time. This has recently been used with benzophenone tetracarboxylic dianhydride (XL) to produce a polymer with good processing ability.[42]

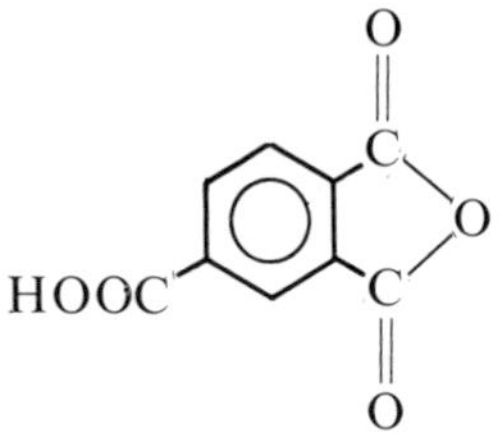

(XLI)

The polymer has been used in foams, compression mouldings and fibres and has good high temperature properties.

Polymers with good processing ability can also be prepared using trimellitic acid anhydride (XLI) to prepare poly(amide-imides).[43]

The polyhydantoins (XLIII) are a series of polymers which have been

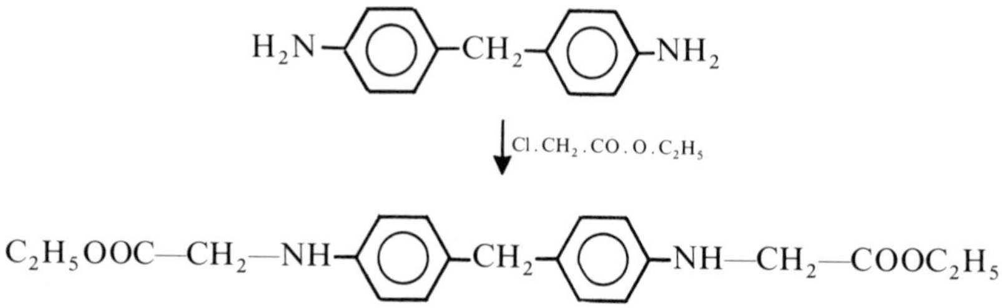

(XLII)

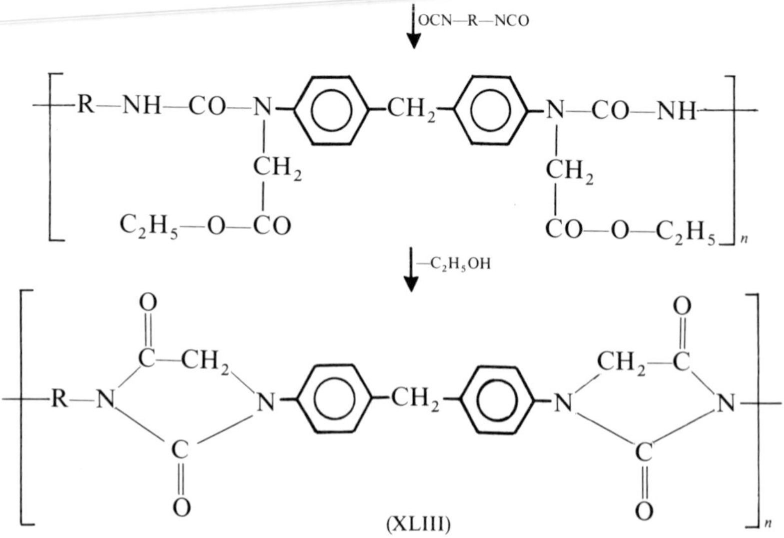

(XLIII)

used for lacquers and electrically insulating varnishes with good high temperature properties. They can be prepared from the reaction of polyglycine esters such as (XLII) with diisocyanates. The compound (XLII) can be prepared by the reaction of aromatic amines with chloracetic ester.[40]

Similar to this synthesis is that of the polyhydrouracils. This involves the reaction of diisocyanates with bis-β-amino esters (XLIV). The latter can be prepared by the addition of acrylic esters to polyamines.[44]

$$NH_2{-}R{-}NH_2 + 2\,CH_2{=}CH{-}COOC_2H_5$$

$$\downarrow$$

$$C_2H_5{-}O{-}CO{-}CH_2{-}CH_2{-}NH{-}R{-}NH{-}CH_2{-}CH_2{-}CO{-}O{-}C_2H_5$$

(XLIV)

$$\downarrow OCN{-}R'{-}NCO$$

$$\left(
\begin{array}{ll}
COOC_2H_5 & COOC_2H_5 \\
| & | \\
CH_2 & CH_2 \\
| & | \\
CH_2 & CH_2 \\
| & | \\
{-}N{-}R{-}N{-}CO{-}NH{-}R'{-}NH{-}CO{-}
\end{array}
\right)_n$$

$$\downarrow {-}C_2H_5OH$$

Finally, the polybenzoxazinediones are worthy of mention. The polymer can be cast from solutions in solvents such as dimethylformamide to form films with good high temperature properties. They are prepared by the reaction of a diisocyanate with a bis-ortho-hydroxyacid ester[45] (e.g. XLV).

CONCLUSION

The major routes for the synthesis of polyurethanes via the two main diisocyanates, tolylene diisocyanate and polymeric MDI, are not likely to be much changed in the near future unless a commercial carbonylation process for TDI is developed.

Many new routes have, however, been developed for the production of new isocyanates which may find speciality applications. Methods are being developed for the *in situ* generation of isocyanates at moderately elevated temperatures which should make it possible to produce one-component heat curable polyurethanes.

Diisocyanates are also finding uses in new high temperature resistant polymers.

REFERENCES

1. Buist, J. M. and Gudgeon, H. (1968). *Advances in Polyurethane Technology*, MacLaren and Sons, Ltd, London.
2. Spitz, P. H. (1973). *Hwahak Konghak (J.KIChE)*, **11**, 125.
3. Marathon Oil Company, US Patent 3,458,448.
4. Twitchet, H. J. (1974). *Chem. Soc. Rev.*, **3**, 209.
5. *Chem. Eng. News*, 24 Oct., 1966, 38.
6. Schmitt, K. (1967). *Angew. Chem. Int. Ed.*, **6**, 375.
7. Campbell, T. W. and Stille, J. K. (1972). Condensation monomers, in *High Polymers*, vol. 27, T. W. Campbell and J. K. Stille, eds., Wiley-Interscience.

8. HARDY, W. B. and BENNETT, R. P. (1967). *Tet. Lett.*, 961.
9. OLIN MATHIESON, British Patent 1,252,517.
10. SHELL, US Patent 3,719,699.
11. Japan Kokai 75, 49,253.
12. ELDERFIELD, R. C. (1961). *Heterocyclic Compounds*, vol. 7, 498, Wiley.
13. CHAPMAN, J. A., CROSBY, J., CUMMINGS, C. A., RENNIE, R. A. C. and PATON, R. M. (1976). *Chem. Comm.*, 240.
14. KLAMANN, D., KOSER, W., WEYERSTAHL, P. and FLIGGE, M. (1965). *Chem. Ber.*, **98**, 1831.
15. ICI, Ger. Offen., 2,336,403.
16. ICI, Ger. Offen., 2,555,830.
17. ICI, Ger. Offen., 2,422,764.
18. ULRICH, H., TUCKER, B. W. and SAYIGH, A. A. R. (1975). *J. Polym. Sci. Polym. Chem. Ed.*, **13**(1), 267.
19. ULRICH, H., TUCKER, B. and SAYIGH, A. A. R. (1966). *J. Org. Chem.*, **31**, 2658.
20. ULRICH, H., SAYIGH, A. A. R. and TILLEY, J. N. (1964). *J. Org. Chem.*, **29**, 2401.
21. SIEFKEN, W. (1949). *Liebigs Ann. Chem.*, **562**, 75.
22. ULRICH, H. and SAYIGH, A. A. R. (1971). *Newer Methods of Preparative Organic Chemistry*, Vol. VI, W. Foerst, ed., Academic Press, New York and London.
23. MCKILLIP, W. J., CLEMENS, L. M. and HAUGHLAND, R. (1967). *Can. J. Chem.*, **45**, 2613; MCKILLIP, W. J. and SLOGEL, R. C., ibid., 2619; SLOGEL, R. C. and BLOMQUIST, A. E., ibid, 2625.
24. HINMAN, R. L. and FLORES, M. C. (1959). *J. Org. Chem.*, **24**, 660.
25. SLOGEL, R. C. (1968). *J. Org. Chem.*, **33**, 1374.
26. CULBERTSON, B. M., SEDOR, E. A. and SLOGEL, R. C. (1968). *Macromolecules*, **1**, 254.
27. *Chem. Eng. News*, 2 Apr., 1973, 11.
28. CULBERTSON, B. M. and RANDEN, N. A. (1971). *J. Appl. Polym. Sci.*, **15**, 2609.
29. MCKILLIP, W. J. (1974). *Adv. Urethane Sci. and Tech.*, **3**, 81.
30. ICI, Ger. Offen. 2,254,611.
31. GITTOS, M. W., DAVIES, R. V., IDDON, B. and SUSCHITZKY, H. (1976). *J.C.S.*, *Perkin I*, 141.
32. SINCLAIR RESEARCH INC., US Patent 3,423,447.
33. SINCLAIR RESEARCH INC., US Patent 3,480,595.
34. ATLANTIC RICHFIELD, US Patent 3,983,958.
35. UNION CARBIDE, US Patent 3,523,924.
36. KAZUO SOGA, SATORU HOGODA and SAKUJI IKEDA (1974). *Die Makromol. Chem.*, **175**, 3309.
37. FARRISSEY, W. J., ALBERINO, L. M. and SAYIGH, A. A. R. (1975). *J. Elastomers Plast.*, **7**(3), 285–314.
38. CAMPBELL, T. W., MONAGLE, J. J. and FOLD, V. S. (1962). *J. Am. Chem. Soc.* **84**, 1493; MONAGLE, J. J., CAMPBELL, T. W. and MCSHANE, H. F., ibid, 4288.
39. SPERANZA, G. P. and PEPPEL, W. S. (1958). *J. Org. Chem.*, **23**, 1922.
40. MERTEN, R. (1971). *Angew. Chem. Int. Ed.*, **10**(5), 294.
41. PATTON, T. L. (1971). *Polym. Preprints*, **12**, 162.
42. CARLETON, P. S. and FARRISSEY, W. J., JR. (1969). *Tet. Lett.*, **40**, 3485; *J. Appl. Polym. Sci.* (1972) **16**, 2983.

43. FREY, H. E., US Patent 3,300,420, 1967: Rochina, V.and Allard, P.,US Patent 3,717,696, 1973.
44. DYER, E. and HARTYLER, J. (1969). *J. Polym. Sci. A*-1, **7**, 833.
45. BOTTENBRUCH, L. (1970). *Angew. Makromol. Chem.*, **13**, 109, 158: Tohyama, S., Kurihara, M., Ikeda, K. and Yoda, N. (1967). *J. Polym. Sci. A*-1, **5**, 2523.

Chapter 3

DEVELOPMENTS IN POLYURETHANE ELASTOMERS

R. P. Redman

ICI Organics Division, Manchester, UK

SUMMARY

Nearly all polyurethane elastomers of importance are block copolymers with a two-phase domain structure. The structure of these polymers is discussed in the first section of this chapter in terms of hydrogen bonding and phase separation effects as well as the thermal history of the polymer. Some structure–property effects are discussed in the second section showing the effect of phase separation and molecular weight on a range of physical properties and also the structure and properties of polyblends made from polyurethane elastomers and other polymers. A final section reviews developments in application technology in the areas of solid reaction moulded polyurethanes, microcellular polyurethanes, thermoplastic polyurethanes and coatings.

INTRODUCTION

Since their commercialisation over 20 years ago, polyurethane elastomers have had a consistently high growth rate. World-wide consumption of solid and microcellular polyurethane elastomers in 1970 was estimated at 35 000 tonnes per year.[1] This had risen to 130 000 tonnes per year by 1975, largely due to the advent of microcellular polyurethanes for shoe soling and for flexible automobile parts, and it is expected to be about 170 000 tonnes per year by 1980.[1]

Matching this growth in the use of polyurethane elastomers has been the growth in the knowledge of the structure of these polymers and the factors which affect their physical properties.

Nearly all polyurethane elastomers of commercial significance are block copolymers and exist as two-phase materials. Recent developments in the understanding of the structure and morphology of these block copolymer elastomers and some aspects of their structure–property relationships are discussed in the first sections of this chapter. A final section looks at some of the more recent developments in the use and application of polyurethane elastomers.

It is impossible to cover all aspects of polyurethane elastomer chemistry and technology in a chapter of this length and emphasis has been placed on those areas which are presently subject to most study and development.

THE STRUCTURE OF POLYURETHANE ELASTOMERS

Intermolecular Interactions and Supramolecular Structure

Elastomeric polyurethane block copolymers consist of alternating blocks of flexible chains of low glass transition temperature (often called 'soft' blocks or segments) and highly polar, relatively rigid blocks ('hard' blocks or segments). The soft blocks are exemplified by aliphatic polyesters and polyethers such as poly(tetramethylene adipate) (PTMA) and poly(oxy-tetramethylene) glycol (PTMG). These have glass transition temperatures (T_g) below room temperature; they usually have low melting points or are amorphous and have molecular weights in the range from about 600 to 3000.

The hard blocks are formed by the reaction of a diisocyanate, such as diphenylmethane diisocyanate (MDI) or tolylene diisocyanate (TDI), with a low molecular weight glycol or diamine such as butane-1,4-diol (BD) or 3,3'-dichloro-4,4'-diaminodiphenylmethane (MOCA). The hard blocks usually constitute about 30–50 % by weight of the total polymer.

The microphase separation of these two dissimilar blocks produces regions of hard block concentration (domains) which act as cross-link points for the soft blocks, thus giving rise to the rubbery nature of these polymers. This microphase separation, and the hydrogen bonded domain structure which results from it, is now recognised as the principal feature controlling the properties of these elastomers.[2,3] However, the degree to which the hydrogen bonding affects the final material properties has been the source of some controversy. Some studies have predicted and shown that the break up of the hydrogen bonding can be a source of mechanical loss,[4,5] while others[6-8] have noted only minimal changes between systems with and without hydrogen bonds. The usual segmented polyurethane systems do have considerable hydrogen bonding capabilities and infra-red spectroscopy has been used to study this.

Seymour et al.[9] concluded that essentially all the NH groups are involved in hydrogen bonding in both polyether (PTMG) and polyester (PTMA) polyurethanes derived from MDI and butane diol. It was estimated that about 60% of the NH groups in the polyether system were associated with the hard block urethane carbonyls (urethane–urethane hydrogen bonding), the rest being associated with the soft block ether oxygens (urethane–soft block hydrogen bonding). Because of several unresolved peaks in the carbonyl region of the spectrum of the polyester polyurethane it was not possible to assess the relative contribution made to the hydrogen bonding in such systems by the two potential acceptors, ester and urethane carbonyl (Fig. 1). However, the degree of phase separation would be expected to be less with the more polar polyester than the polyether and this would give rise to a lower level of urethane–urethane bonding than the 60% observed with the polyether.

Other similar studies on polyurethane[10] and poly(urethane-urea)[11] block copolymers support this view of incomplete segregation of the blocks. These general conclusions are further reinforced by many studies of the thermal and thermo-mechanical behaviour of polyurethane block copolymers.[12] Differential scanning calorimetry (DSC) has been the most widely used technique, and generally the observed transitions fall into three main groups: those below $-30°C$ associated with the glass transition temperature of the soft block, transitions in the region $80°C$ to $150°C$ and those above $160°C$ associated with the thermal dissociation of the hard block aggregates which may be crystalline or paracrystalline.

The endotherms occurring in the region of $80°C$ have frequently been ascribed to the dissociation of the urethane–soft block hydrogen bonds and those in the region of $150°C$ to the break-up of urethane hydrogen bonds.[13-15]

Seymour,[16] however, reports that on annealing the elastomer the transition at $80°C$ can be progressively moved to higher temperatures until it merges with the transition of $150°C$. At higher annealing temperatures this transition is itself moved and it eventually merges with the transition above $160°C$ associated with the break-up of the hard block aggregates. Interpretation of these results in terms of hydrogen bond dissociation would require that no dissociation occurred in a well-annealed sample below the transition at $160°C$. This is in contrast to the actual hydrogen

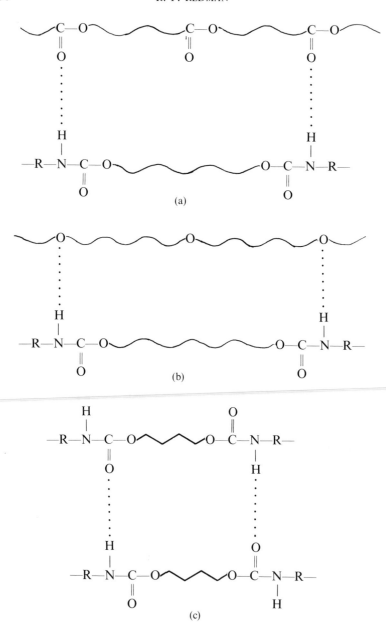

FIG. 1.   Diagrammatic representation of hydrogen bonding. (a) Ester–urethane, (b) ether–urethane, (c) urethane–urethane.

bond breakdown behaviour as shown by infra-red spectroscopy. The level of hydrogen bonding progressively decreases with the increasing temperature so long as the temperature is above the T_g of the hard block (about 80 °C) and even at 200 °C there is still some hydrogen bonding present. Seymour therefore suggests that the endotherms observed in the DSC traces can be attributed to the loss of long and short range order. Different degrees of short range order may exist simultaneously due to the distribution of hard block lengths. The short range ordering may be continuously improved by annealing as shown by the merging of the endotherms. The thermal behaviour of the hydrogen bonds is insensitive to the degree of ordering present and is affected primarily by the T_g of the hard block. Hydrogen bond dissociation only occurs above the T_g of the hard block. For pure MDI–butane diol homopolymer this T_g is reported to be 109 °C,[17] but because of the short block lengths in polyurethane copolymers it is usually somewhat lower than this (80 °C for the samples examined by Seymour) and for a given chemical composition is very dependent on the hard block length. It is suggested[16] that hydrogen bonding is less important than has generally been thought. Chain mobility controls hydrogen bond dissociation rather than vice versa and the unusually good mechanical properties of polyurethanes stem from the incompatibility of the blocks and the consequent phase separation, rather than from the presence of hydrogen bonds *per se*. The presence of hydrogen bonds serves to increase the overall cohesion as these are stronger and more directional bonds than other intermolecular forces.

Hydrogen bonding may be more important with highly crystalline polyurethanes where interchain alignments giving the greatest degree of hydrogen bonding are favoured, as shown by Bonart[2] in an X-ray study of elastomers with MDI–ethylene diamine and MDI–hydrazine hard blocks.

Bonart has also considered the consequences of hydrogen bonding in the hard block aggregates.[18] The mutual hard block affinity would be expected to increase with an increasing number of stress-free hydrogen bonds per urethane block and this would be reflected in the heat distortion temperature (HDT) of the elastomers. Bonart considered two-dimensional models of MDI–butane diol and MDI–pentane diol hard block systems. (Simplified structures are shown in Fig. 2.) From these models it would have been expected that the pentane diol based systems would be more stable (more stress-free hydrogen bonds per urethane block) and therefore have a higher HDT than the butane diol based system. This is contrary to observation and Bonart suggests a three-dimensional model to allow completion of the urethane–urethane hydrogen bonding. This gives rise to

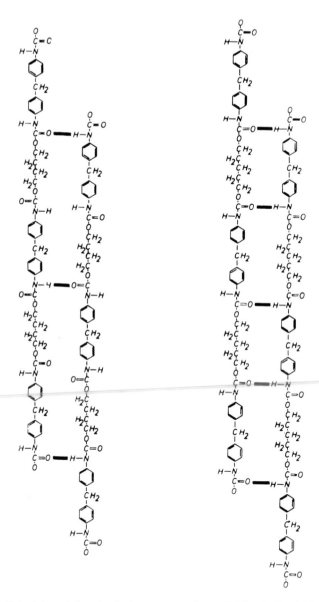

FIG. 2.   Principles of the physical structure of crosslinking in hard blocks with butane or pentane diol as chain extender. Left: butane diol as chain extender; right: pentane diol as chain extender.[18] Reproduced by courtesy of Marcel Dekker Inc., New York.

the concept of transverse cross-linking by urethane groups acting in different directions (Fig. 3). Figure 3 represents only two cross-linking planes. A large number of additional cross-linking planes (not shown) also have to be taken into account so that each hard block within a hard block domain is found at the intersection of two such planes. Bonart has also considered models of diamine extended polyurethanes. In these cases excess NH groups are present, in contrast to the diol extended polyurethanes. Since no free NH groups are seen in the infra-red spectra of diamine extended polyurethanes, one carbonyl must be able to associate with the two NH groups simultaneously.[19] These NH groups interact with carbonyl groups above or below their plane thus causing transverse cross-linking.

This may provide an explanation for the absence of segregation and elasticity in diol extended polyurethanes below a molar ratio of 1:3:2 of soft block:MDI:diol. Such segregation and elasticity is found, however, in amine extended systems down to molar ratios of 1:2:1 since NH groups are available for transverse cross-linking.

This model, although it has recently been questioned,[20] does lead to a consideration of the larger scale structure in these polymers. Bonart's model naturally leads to a lamellar arrangement of the hard blocks, the lamellae growing by lateral accumulation of hard block units, but limited in the thickness dimension to the average hard block length. This gives rise to the spherulitic structure often observed in polyurethane block copolymers. Wilkes[21] has also considered the larger scale orientation and spherulitic structure in crystalline polyurethanes. It was shown that polyurethane elastomers based on poly(oxytetramethylene) glycol, MDI and piperazine cast from solution contain coalesced spherulitic entities. These spherulitic units incorporate both the soft and hard blocks and two models of the structure were proposed (Fig. 4). In the first (Fig. 4a) the hard blocks are radially orientated and hard block lamellae extend tangentially, and in the second the hard blocks are tangentially orientated and the lamellae grow radially (Fig. 4b). On the basis of positive birefringence it was concluded that the first model with radial hard block orientation was being observed.

Schneider et al.[20] have examined crystalline polyurethanes based on ethylene oxide tipped poly(oxypropylene) glycol polyethers, MDI and butane diol and have shown the spherulites to be made up of fibrils or strands. They suggested that the fibrils consisted of stacks of soft and hard blocks, the spacing being affected by the soft block length.

Since information on crystallographic order cannot be transmitted through the soft block phase, continued crystal growth in the hard block

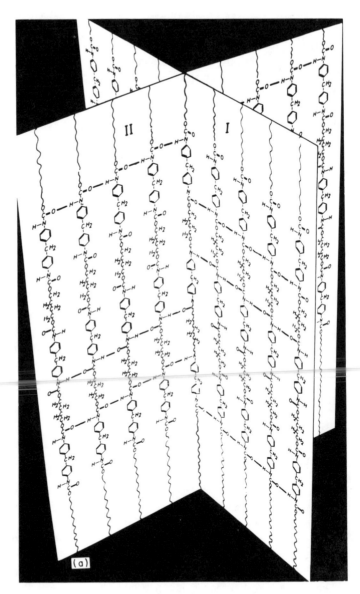

FIG. 3.   Diagrammatic representation of the three-dimensional crosslinking structure. (a) Extended with butane diol; (b) extended with pentane diol.[18] Reproduced by courtesy of Marcel Dekker Inc., New York.

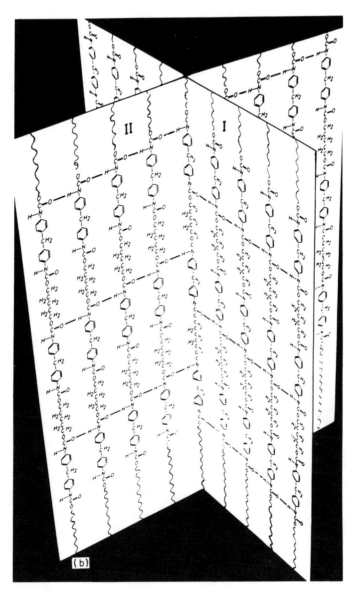

FIG. 3.—contd.

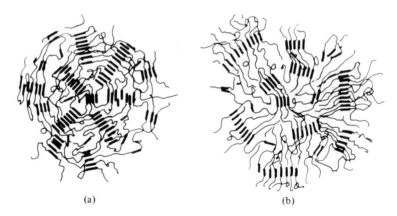

(a)   (b)

FIG. 4. Possible models of spherulitic structure. Assuming the principal polarisability axis lies along the chain backbone, model (a) will be negatively birefringent while model (b) will be positive.[21] Reproduced by courtesy of John Wiley Inc., New York.

direction requires repeated primary nucleation. This is a slow process compared to addition of hard blocks to the lateral crystal surfaces and the hard blocks are therefore expected to lie perpendicular to the fibril axis. This conclusion is in contrast to Wilkes' proposed model and Schneider suggests that shown in Fig. 5.

The transverse orientation of the hard blocks in the model is further supported by the observations of Seymour *et al.*[22] with both polyether and polyester based elastomers and Kimura *et al.*[23] with polyether based systems using infra-red dichroism studies. They have studied the orientation of the hard blocks at low elongations using the NH and carbonyl stretching frequencies. In both cases the initial transverse orientation to the stretch direction was regarded as being due to the rotation of the long axis of lamellae crystallites into the stretch direction. At high elongations a change to parallel hard block orientation was brought about by a disruption of the original lamellae and a reorganisation of the hard blocks as proposed by Bonart.[2]

The morphology and structure of polyurethane block copolymers has been recently reviewed in some detail[24] and present views can be summarised as follows:

1. Because the hard and soft blocks are partly incompatible with each other, the elastomers show a two-phase morphology although there is a significant level of mixing of the hard and soft blocks.

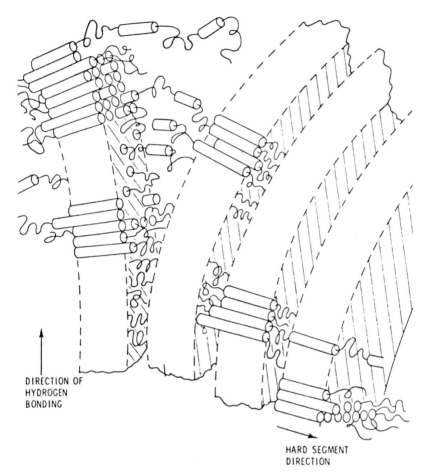

DIRECTION OF
HYDROGEN
BONDING

HARD SEGMENT
DIRECTION

Fɪɢ. 5. Suggested organisation of crystalline hard segment strands. The average orientation of the strands is in the radial direction.[20] Reproduced by courtesy of Marcel Dekker Inc., New York.

2. Hydrogen bonding can occur between hard and soft blocks although the extent to which this is responsible for physical properties is not certain.

3. Hydrogen bonding occurs between individual hard blocks giving rise to a three-dimensional molecular domain structure.

4. These domains may themselves be in a larger, ordered arrangement including both soft and hard blocks, the hard blocks being built up

in a transverse orientation to their molecular axis leading, in cases, to the appearance of spherulites in the polymer.

5. The morphology is unstable with respect to temperature and is dependent on both the chemical constitution and thermal history of the polymer.

Temperature and Time-Dependent Effects

As was indicated in the previous section, the domain structure of polyurethane block copolymer elastomers is an unstable morphology with respect to temperature. For example, it has been shown that annealing causes a shift in the DSC endotherm at 80 °C until it eventually merges with the endotherm at 150 °C and this was attributed to an increase in the short range order.[16]

On the other hand Wilkes[25] has reported on the basis of stress–strain behaviour, T_g[26] and small angle X-ray[27] measurements that, upon heating, the degree of the domain formation apparently decreases so that more mixing occurs between hard and soft blocks. The area is clearly complex and two quite distinct parameters need to be borne in mind: (a) the level of phase separation and hard block–soft block interaction and (b) the degree of ordering and crystallinity within the hard block domains.

Wilkes has concentrated on the former, the degree of phase separation, and this is clearly temperature sensitive. The degree of mixing of two phases generally increases with the temperature and by the theory of elasticity[28] the retractive forces on the soft blocks would also be expected to increase with temperature, tending to 'pull' hard blocks from the domains. Both factors would tend to increase compatibility as the temperature is raised and an equilibrium state of phase separation would be reached dependent on the temperature and chemical constitution. Due to the limited mobility of polymer chains, particularly where they are restricted by the hard blocks, the equilibrium position may take some time to attain. On cooling the original structure is reformed. This behaviour is illustrated in the model shown in Fig. 6.

Wilkes[26] has studied the effect of time on the structure of polyurethanes after annealing at 170 °C for 5 min. The T_g of both polyether and polyester based systems decreased with time over a period of several hours at 25 °C as the original domain structure was reformed and phase separation increased. This is because the T_g of the soft block is sensitive to the degree of phase separation and rises as the mobility of the soft blocks is restricted by the hard blocks as the phases become mixed, and falls as they separate (see next section). Similarly, the modulus of the systems increased with time after annealing as the reinforcing effect of the domain structure became established and this was reflected in the stress–strain behaviour.

It was also shown that polyether systems are much quicker to reach equilibrium than the equivalent polyester based systems, probably due to the greater intermolecular forces in the polyester system and also the lower T_g, and therefore greater mobility, of the polyether chains.

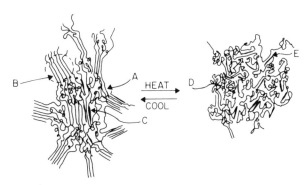

FIG. 6. Schematic model depicting the morphology for both (i) long time (control) and (ii) following heat treatment. A = partially extended soft block; B = hard-block domain; C = hard block; D = cooled or 'relaxed' soft block; E = lower-order hard-block domain.[26] Reproduced by courtesy of the American Institute of Physics.

As well as these effects on phase separation, annealing also alters the ordering within the hard domains. If annealing is carried out above the T_g of the hard block but below its melting point, then there is sufficient mobility within the hard blocks for reordering to occur. Due to the ability of these blocks to crystallise, this reordering results in improved structure and selective crystallisation. Thus while there may be fewer hard domain aggregates, those that are present tend to be better ordered and have a higher degree of crystallinity.

This is essentially the behaviour noted by Seymour[16] with the merging of the DSC endotherms.

This structuring of the hard domains can be clearly seen in the DSC endotherm associated with the melting process, T_m. The T_m of both polyether and polyester based elastomer systems moves to a higher temperature and becomes narrower and better defined by annealing at 200 °C or 210 °C.[28] This behaviour follows a typical rate process showing a rapid change initially which levels off with time as the crystallinity approaches that of an MDI–butane diol homopolymer with a T_m of

244–245 °C. This upward shift in T_m occurs simultaneously with the upward shift in the soft block T_g indicating that both phase mixing and hard block structuring are occurring at the same time.

An explanation of the absence of any improved structure or crystallinity in Wilkes' X-ray study of annealed elastomers may be due to a rate effect. The rate of hard block ordering is probably slower than the mixing of the phases, particularly at a temperature of 170 °C as used by Wilkes, owing to the greater mobility of the soft blocks.

If the temperature is taken above the melting point of the hard blocks then complete loss of structure occurs as shown by DSC. This behaviour is illustrated in Table 1 for a system based on a 2000 molecular weight polyester, MDI–butane diol.[29]

TABLE 1

DSC ENDOTHERM BEHAVIOUR FOR A 2000 MW POLYESTER BASED POLYURETHANE WITH MDI–BUTANE DIOL HARD BLOCKS BEFORE AND AFTER ANNEALING

Annealing conditions	T_g °C	T_m °C
Control	−30	211
4 h at 210 °C	−24	237
After melting at 280 °C and immediate re-run	−16	Broad endotherm centred at 163

The structuring observed on annealing below the T_m is frozen as the elastomer is cooled again below the T_g of the hard blocks (usually 80–100 °C). It seems likely that superimposed on this structure is a lower order of structure due to the aggregation of hard blocks from the mixed soft and hard phases. This phase separation is only frozen below the T_g of the soft blocks. In contrast to the mixing of the phases by annealing, therefore, the ordering and selective crystallisation of the hard blocks is not entirely reversible.

STRUCTURE–PROPERTY EFFECTS IN POLYURETHANE ELASTOMERS

Structure and Phase Separation Effects

The dependence of the structure–property relationships of polyurethane block copolymer elastomers on their chemical composition (nature of the polyether or polyester soft block, type of diisocyanate and chain extender, etc.) and principal physical properties (strength, hydrolytic stability, etc.) are now well recognised and have recently been extensively reviewed.[12] The mechanical properties of these polyurethane elastomers have been discussed in terms of chain flexibility,[3] rigidity of the aromatic units[30,31] and the degree of interchain interaction.[12]

In the last few years there has been a growing appreciation that the physical properties are determined not only by the chemical structure but also by the extent of the phase separation between the soft and the hard blocks.[16] The two factors are, clearly, closely related and a change in the chemical structure invariably alters the degree of phase separation.

The molecular feature most affected by the level of phase separation is the mobility of the polymer chains. If the phases are completely separated then the mobility of the flexible chains is determined by the T_g of the soft block polymer and the elastomer would show two distinct T_gs associated with the soft and with the hard blocks, respectively. At higher temperatures the polymer would show a narrow, well-defined melting point associated with the break-up of the hard blocks. At the opposite extreme, if the two blocks are completely compatible, then a single broad T_g and a broad, ill-defined melting range would be observed.

Most polyurethanes lie between these two extremes, as was indicated in the previous section, and are dependent on hard block–soft block interactions for many of their property features.

As the level of phase separation decreases and the compatibility increases, the hard blocks play an increasingly important part in restricting the mobility of the soft blocks, and the soft blocks in disrupting the ordering in the hard block domains.

Some properties of polyurethane block copolymers which are therefore sensitive to the degree of phase separation are the low and high temperature properties, the modulus–temperature behaviour and the hysteresis and related properties.

Low and High Temperature Properties

The low temperature behaviour of polyurethane elastomers is primarily determined by the T_g of the soft blocks. This is influenced not only by the nature of the soft block (e.g. polyethers generally have a lower T_g than polyesters) but by the level of phase separation between the hard and soft blocks.

In a study of polycaprolactone based elastomers, Seefried *et al.*[32] showed that the T_g of the soft block shifted to higher temperatures as the molecular

TABLE 2
EFFECT OF VARIOUS PARAMETERS ON T_g IN POLYCAPROLACTONE SYSTEMS

(a) *Effect of MW of polycaprolactone on T_g in PCL/MDI/BD, 1/2/1 (average hard block MW = 590)*

MW PCL	340	530	830	1 250	2 100	3 130
T_g °C	53	25	−10	−27	−40	−45

(b) *Effect of hard block length on T_g with PCL MW 2 100 in PCL/MDI/BD*

Molar ratio PCL/MDI/BD	1/2/1	1/3/2	1/4/3	1/5/4	1/6/5
Average MW hard block	590	930	1 270	1 610	1 950
T_g °C	−40	−40	−32	−30	−30

(c) *Effect of hard block length on T_g with PCL MW 830 in PCL/MDI/BD*

Molar ratio PCL/MDI/BD	1/2/1	1/3/2	1/5/4	1/8/7
Average MW hard block	590	930	1 610	2 630
T_g °C	−10	8	24	50

(d) *Effect of MDI and TDI based hard block with PCL MW 2 100*

Molar ratio PCL/NCO/BD	1/4/3	1/6/5	1/8/7	1/4/3	1/6/5	1/8/7
Average MW hard block	1 270	1 950	2 630	966	1 494	2 022
T_g °C	−32	−30	−30	−20	0	10
	MDI based			TDI based		

(e) *Effect of HBPA chain extender on T_g with PCL 2 100 in PCL/TDI/HBPA*

Molar ratio PCL/TDI/HBPA	1/2/1	1/4/3	1/6/5	1/10/9
Average MW hard block	584	1 404	2 224	3 864
T_g °C	5	12	22	55

weight of the polycaprolactone decreased. This is indicative of the restriction on the mobility of the soft block by the rigid hard blocks as the compatibility increases at the lower molecular weights (Table 2a). If the molecular weight of the polycaprolactone was above about 2000, then the T_g was fairly insensitive to the hard block length[33] indicating that the phases are well separated (Table 2b). However, as the molecular weight of the polycaprolactone was reduced, phase separation decreased and the T_g became sensitive to the hard block length (Table 2c).

These studies were made with a system containing MDI–butane diol hard blocks and these rigid blocks enhance phase separation. Similar studies based on commercial TDI (a mixture of 2,4 and 2,6 isomers) showed much less perfect domain formation[34] and greater compatibility of the phases. The T_g of the soft blocks was therefore very dependent on the hard block length (Table 2d), the T_g increasing with increasing block length.

Changing from butane diol to 2,2-bis(4-hydroxycyclohexyl)propane (HBPA) as chain extender in these TDI based elastomers resulted in polymers with only very low, or possibly no phase separation.[35] The T_g was therefore very dependent on the hard block length (Table 2e).

Schneider[36] has made a similar study of polyether based polyurethane elastomers, where phase separation is normally more pronounced than in polyester types, and noted a similar upward shift of the T_g from that of the polyether homopolymer due to the restriction of mobility of the soft blocks by the hard domains. However, at very low hard block lengths (1:2·2:1 molar ratio of polyether:MDI:BD, with a molecular weight of the hard blocks of about 600) the T_g dropped to that of the soft block homopolymer suggesting that there is a minimum length of hard block necessary to restrict the mobility of the chains.

The effect of different diisocyanates on the T_g of poly(ethylene adipate) and butane diol based elastomers has recently been reported.[37] The isocyanates used were MDI, TDI (mixed 2,4 and 2,6 isomers), HDI (hexamethylene-1,6-diisocyanate), H_{12}MDI (bis(4-isocyanatocyclohexyl)-methane) and IPDI (isophorone diisocyanate).

Again the T_g of the TDI based system ($-16\,°C$) was higher than the MDI based system ($-31\,°C$), indicative of reduced phase separation in the former system. However, the T_gs of the aliphatic diisocyanate based systems were lower than the MDI control (HDI, $-42\,°C$; H_{12}MDI, $-39\,°C$ and IPDI, $-32\,°C$ to $-35\,°C$) suggesting increased phase separation and this was attributed to the stronger hydrogen bonding in the hard block domain with the aliphatic diisocyanates.

At the other end of the scale, the processing temperature of thermoplastic polyurethane elastomers can be affected by phase separation factors. If the phases are well-separated, then the melting point of the polymers depends on the hard block length and the hard block crystallinity and is usually characterised by a narrow, well-defined endotherm in a DSC trace. By decreasing the phase separation, the melting range of the polymers can be significantly broadened and sometimes lowered. This effect can be considered as a plasticisation of the hard block domains by the soft blocks.

This effect readily can be seen by comparing the DSC traces of thermoplastic polyurethanes based on 1000 and 2000 molecular weight poly(tetramethylene adipate) soft blocks. The former system shows only a small broad endotherm centred at 150°C, indicative of the break-up of the hard blocks over a wide temperature range. By contrast, the system based on the 2000 molecular weight soft blocks shows a well-defined, narrow endotherm at 202°C, showing the more independent behaviour of the soft and hard blocks.[38] The greater compatibility of the 1000 molecular weight PTMA system is further evidenced by the higher T_g of the soft block ($-38\,°C$) compared to the 2000 molecular weight PTMA system ($-48\,°C$).[39] Phase separation factors can therefore be important in the processing of thermoplastic polyurethanes.

Broader melt processing temperature ranges have been obtained by making a prepolymer from a polyester and TDI using an excess of the polyester.[40] This urethane extended polyester is subsequently reacted with the MDI and butane diol in the normal manner to give a polymer containing TDI based urethane groups as well as MDI based urethane groups. This effectively reduces the level of phase separation and consequently gives a broader processing temperature range.

Modulus–Temperature Effects

The modulus of an elastomer is partly determined by the mobility of its component chains. A polyurethane block copolymer with a high degree of phase separation would therefore be expected to show two distinct drops in modulus, as the temperature is increased, associated with the onset of mobility (T_g) in the soft and hard blocks. A theoretical modulus–temperature curve would therefore be as in Fig. 7a with a level plateau between the two dramatic falls in modulus. On the other hand, if the two phases are completely compatible, then a single

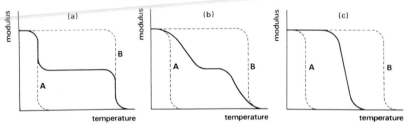

FIG. 7. Modulus–temperature behaviour of block copolymers. (a) = complete phase separation; (b) = phase separation but with some mixing; (c) = compatible phases.[44] Reproduced by courtesy of Kolloid Gesellschaft.

fall in modulus over a fairly narrow temperature range would be expected at the T_g of the combined blocks (Fig. 7c). In reality the difference is not so marked as these two curves. A typical curve for a polyurethane elastomer is shown in Fig. 7b.

However, by altering the degree of phase separation, it is possible to alter the shape of the modulus–temperature curve. This has been extensively studied in the last few years in connection with polyurethane elastomers for flexible automobile parts. Effort has been directed towards developing a system where the modulus changes as little as possible with temperature. This has been done by optimising phase separation by control of the molecular weight of the polyol, the hard block length, the chain extender composition, branching in the hard blocks[41,42] and by the use of a mixture of incompatible polyols of different molecular weight.[43] This topic is covered in further detail in Chapters 5 and 9.

Hysteresis and Related Properties

Polyurethane rubbers are noted for their generally high hysteresis loss.[12] Much of the hysteresis loss occurring during the extension and subsequent relaxation of an elastomeric polyurethane block copolymer is due to the orientation and restructuring of the molecular organisation initially present.[2] Although at high stresses much of this restructuring is in the hard block domains,[12] at lower levels of stress the restructuring is primarily between hard and soft blocks. Such restructuring and transient 'couplings' are therefore very dependent on the degree of hard block–soft block interactions and hence on the level of phase separation. Generally, the higher the level of phase separation the lower the hysteresis loss. Thus, polyether based polyurethanes generally have a lower hysteresis loss than the corresponding polyester based elastomers.[45]

Similarly, the hysteresis loss of an elastomer based on 1000 molecular weight poly(tetramethylene adipate), MDI and butane diol is higher than the corresponding elastomer prepared from 2000 molecular weight poly(tetramethylene adipate) at the same hard block content.[38] The hysteresis loss (on the third cycle) for the two systems at 50% extension is shown in Table 3 for four different isocyanate:hydroxyl ratios.

Properties very much related to hysteresis are the tear strength, the cut growth[46] and the resilience of elastomers.

TABLE 3

HYSTERESIS LOSS (%) ON THE THIRD CYCLE AT 50% ELONGATION FOR PTMA/MDI/BD ELASTOMERS AT VARIOUS ISOCYANATE:HYDROXYL RATIOS (35% Hard Block Content)

NCO:OH ratio	0·96	0·98	1·00	1·02
1 000 MW PTMA	28	26	17	24
2 000 MW PTMA	24	17	13	21

The use of a mixture of polyethers or a mixture of polyesters of different molecular weight to improve the tear strength and cut growth properties has been disclosed.[47-50] This improvement may be as a result of changes in the degree of phase separation caused by the use of soft blocks having two molecular weight peaks in their distribution curves.

The use of mixed isocyanates[40] to obtain a leather-like feel, rather than a snappy feel, to urethane films has also been described and this is undoubtedly due to a decreased level of phase separation in the system.

Molecular Weight Effects in Thermoplastic Polyurethanes

Although thermoplastic polyurethanes elastomers have been available since 1958, little has been published on the effect of the molecular weight of the polymers on their physical properties.

The molecular weight of polyurethane block copolymer elastomers can be varied by either altering the overall isocyanate:hydroxyl ratio, the molecular weight falling at isocyanate ratios below unity, or by the addition of a monofunctional alcohol chain stopper. At isocyanate ratios above equivalence, allophanate cross-linking is possible,[51] but at low levels of cross-linking the thermoplasticity of the polymers is not affected, allophanate links being reversibly broken at temperatures below the processing temperature.[52]

In a recent study[53] Schollenberger examined the effect of the molecular weight on a range of physical properties. The elastomers were based on poly(tetramethylene adipate) of molecular weight 1110, MDI and butane diol. The polymers were made in solution, the molecular weight being controlled by the addition of propanol when the desired degree of polymerisation had been reached as determined by the solution viscosity.

It was found that many of the properties increased with molecular weight (number average molecular weight, $\bar{M}_n$) and then levelled off at a value of $\bar{M}_n$ of 35 000–40 000. This type of behaviour was noted for the intrinsic viscosity, specific gravity, processing temperature, tensile strength, 300 % modulus, abrasion resistance, Clash–Berg rigidity modulus, Gehman freeze point and stick temperature.

Other polymer properties increased steadily with increasing $\bar{M}_n$ without any indication of levelling off. These included the Brookfield viscosity and stress relaxation.

Polymer properties decreasing with increasing $\bar{M}_n$ included the melt index, elongation at break and also, more surprisingly, the flex life, film tear strength, hysteresis and modulus at 5 % and 30 % extensions. The hardness, T_g and abrasion showed only random behaviour.

In a further study[38] a comparison was made between elastomers of various molecular weights based on poly(tetramethylene adipate) of molecular weight 1000 and 2000, MDI and butane diol. The hard block content was maintained at 35 % and the molecular weight in the two series altered by preparing the polymers at six different isocyanate:hydroxyl ratios (0·92, 0·94, 0·96, 0·98, 1·00 and 1·02).

The effects on molecular weight (weight average molecular weight, $\bar{M}_w$, and number average molecular weight $\bar{M}_n$) and on rheological, thermal and mechanical properties of varying the ratio of isocyanate:hydroxyl are summarised in Tables 4 and 5.

It can be seen that at low molecular weights the melt viscosities of the polymers based on PTMA of molecular weight 1000 are lower than those based on PTMA of molecular weight 2000. This is probably due to the shorter urethane hard blocks (at the same overall hard block content) in the former systems. This gives rise to less phase separation and consequently less melt reinforcement. This reduced phase separation is shown by the higher T_g (8–11 °C higher) in these systems at a given isocyanate:hydroxyl ratio.

For both systems the T_g rises at high isocyanate:hydroxyl ratios, probably due to reduced chain end effects.

The tensile and tear strengths generally increase and reach a plateau region with increasing molecular weight, although the plateau area does not seem to have been reached in the series based on PTMA of molecular weight 2000.

It is apparent that the molecular weight of these polymers is important in determining many processing and end-use properties and several patents have been published describing methods of controlling the chain lengthening reaction. The use of crash cooling of the polymerising mass[54] and the addition of mono alcohols[55] have been disclosed as well as the use of secondary alcohols[56] to minimise allophanate formation near equivalence of isocyanate:hydroxyl.

Polyblends of Polyurethanes

In recent years there has been growing interest in the modification of the properties of polyurethane elastomers by the use of blends of polyurethane with other polymers.[57–59] Since mixtures of polymers are usually incompatible and will macrophase separate, special methods have to be employed for the synthesis of useful polyblends.

There are essentially four different theoretical approaches to making polyblends: (a) the simultaneous formation of the two polymer systems, the

TABLE 4

EFFECT OF NCO/OH RATIO ON MOLECULAR WEIGHT AND MELT VISCOSITY OF POLYURETHANE BLOCK COPOLYMERS (PTMA/MDI/BD; 35% HARD BLOCK CONTENT)

NCO:OH	$\bar{M}_w$ ($\times 10^{-4}$)	$\bar{M}_n$ ($\times 10^{-4}$)	$\bar{M}_w/\bar{M}_n$	Peak MW ($\times 10^{-4}$)	Viscosity (Nsm^{-2}) at 180°C		
					$10\,s^{-1}$	$100\,s^{-1}$	$1\,000\,s^{-1}$
Based on 1 000 MW PTMA							
0·92	4·96	2·05	2·4	3·84	210[a]	200	105
0·94	6·69	2·51	2·7	4·83	320[a]	295	165
0·96	10·3	3·95	2·6	7·67	1 100[a]	720	300
0·98	17·4	4·56	3·8	12·0	4 400	1 700	490
1·00	36·0	7·83	4·6	18·0	13 800	3 806	880
Based on 2 000 MW PTMA							
0·92	6·17	1·34	4·6	4·83	2 300	660	190
0·94	6·70	1·31	5·1	5·42	2 600	800	240
0·96	9·00	2·27	4·0	6·84	4 800	1 500	370
0·98	15·1	5·22	2·9	12·0	6 000	2 200	550
1·00	19·4	2·89	6·7	14·8	7 600	2 750	650
1·02	25·6	5·88	4·4	18·0	13 300	4 250	1 020

[a] Extrapolated.

TABLE 5

EFFECT OF NCO/OH RATIO ON THERMAL AND MECHANICAL PROPERTIES OF POLYURETHANE BLOCK COPOLYMERS (PTMA/MDI/BD; 35% HARD BLOCK CONTENT)

	NCO:OH	T_g (°C)	T_m (°C)	Comments	Hardness (Shore A)	Tensile strength ($MN\,m^{-2}$)	Elongation (%)	Trouser tear (Nm^{-1} $\times 10^{-3}$)
Based on 1 000 MW PTMA	0·92	−39	174–185	Small, poorly defined endotherms	84	12·4	650	42
	0·94	−36	162		83	17·0	620	45
	0·96	−38	156		84	29·3	600	50
	0·98	−37	165		84	36·2	570	54
	1·00	−34	219		83	37·9	540	51
	1·02	−33	219		83	34·7	520	53
Based on 2 000 MW PTMA	0·92	−47	212	Strong endotherms, also small endotherms near 31–34°C	85	15·8	650	42
	0·94	−47	197–212		85	17·2	600	41
	0·96	−48	202		84	22·9	580	57
	0·98	−47	190		84	32·2	540	57
	1·00	−44	175	Small endotherms	83	36·2	530	53
	1·02	−41	175		83	39·1	510	58

polymers being chosen so that they entangle together and do not separate *en masse*; (b) the formation of the polyurethane followed in a second step by the formation of the blend polymer such that polymerisation takes place within the polyurethane matrix and phase separation is limited to domains within the interstices; (c) the formation of the blend polymer followed in a second step by the formation of the polyurethane, the blend polymer being present as a solution, or more usefully as a dispersion, in the polyurethane precursors; and (d) the blending and mixing of the two preformed polymers by thermoplastic processing or by the mixing of solutions, etc. All four methods have found application in the synthesis of polyurethane polyblends.

(a) *The Simultaneous Formation of a Polyurethane and a Blend Polymer*
This method is very limited in its applicability due to the need to choose a blend polymer which is formed at the same rate as the polyurethane and by a mechanism which does not interfere with the polyurethane cure.

Frisch *et al.*[60-62] have made a study of polyurethane–polyacrylate 'simultaneous interpenetrating networks' (SINs) and shown that the tensile strength of such composites can be higher than that of either of the two individual components, although the particular polyurethane elastomer chosen was a low strength type. This increase in tensile strength was attributed to additional entanglements and interpenetration of the two networks. The composites were clear and only a single T_g was detected indicative of no phase separation.[63]

Frisch has also examined SINs composed of polyurethane –polyepoxide[64] and polyurethane–polystyrene.[65] In this latter case, phase separation was shown to have occurred by the appearance of a domain structure within the matrix in electron micrographs of the samples and also by the presence of two T_g endotherms associated with the two individual polymers. The T_gs were, however, shifted closer together from that of the homopolymers suggesting some mixing at the phase boundaries.

In these systems the word 'simultaneous' is probably used somewhat loosely since it is unlikely that both polymerisation reactions do proceed together and one would be expected to be in advance of the other. This could have a crucial effect on the final morphology if phase separation does occur.

Similar SINs made from polyurethanes and poly(methylmethacrylate) or polystyrene[66,67] showed a two-phase domain structure.

Other polyblend systems which might be considered a simultaneous polymerisation are the latex interpenetrating networks, LIPNs,[68-71] formed by mixing latices of two different linear polymers together with cross-linking agents and subsequently curing by fusing the particles and cross-linking them *in situ*.

Klempner[70,71] has studied poly(urethane-urea)–polyacrylate LIPNs. Dynamic mechanical relaxation measurements showed two T_gs corresponding to the two networks suggesting phase separation. DSC, on the other hand, showed a single, very broad T_g. It was suggested[58] that mechanical relaxation results from motion in the chain segments while the DSC corresponds to a larger scale motion of whole molecules or groups of molecules. In other words, a single T_g in the viscoelastic spectrum indicates total segmental interpenetration while a single T_g in the DSC spectrum shows mixing at the levels of small clusters of molecules or domains.

This touches upon the whole question of the definition of compatibility. Chemical compatibility implies complete molecular mixing but with large polymer molecules compatibility is a relative term. In a recent paper, Kaplan[72] discusses the relevance of T_g to compatibility and interpenetration and defines a compatibility number N_c. In a compatible system with one T_g, $N_c \to \infty$ and in an incompatible system with two T_gs, $N_c \to 0$. The point at which semi-compatibility occurs ($N_c = 1$) is taken as the approximate segment length associated with a T_g, and Kaplan suggests a length of 150 Å. This approximates to 100–5000 carbon–carbon bonds for an associated T_g.

(b) The Sequential Formation of the Polyurethane Followed by the Blend Polymer

This method gives greater control over the morphology of the polyblend as the phase separation of the blend polymer is limited by the constraints imposed by the polyurethane matrix and the blend polymer therefore forms domains within the interstices of the polyurethane.

This interstitial polymerisation technique has been extensively studied by Allen *et al.*[73-77] using vinyl-modified amorphous polyurethane elastomers. The blends were prepared by forming the polyurethane in the presence of a vinyl monomer plus initiator to give a network swollen with monomer. The subsequent polymerisation of the vinyl monomer by raising the temperature gave rise to the two-phase system. It was shown that even at high polyvinyl levels the polyurethane phase was continuous and the polyvinyl present as domains in the interstices of the matrix.

This work on the interstitial polymerisation method has been extended to a study of polyvinyl-modified polyurethane block copolymer elastomers.[78]

By the use of a glassy polyvinyl phase, substantial reinforcement of the polyurethane elastomer was obtained.[79] The polyvinyl phase increased the hardness of the elastomers and the mechanical properties of elastomers containing under 25 % by weight of hard block could be enhanced to and even surpass those of normal polyurethane block copolymers of similar hardness. It was shown that the properties of the polyblend were dependent on the level of cross-linking in the vinyl phase as well as the nature of the polyvinyl. Polar species like poly(methyl methacrylate) gave rise to substantial mixing of the phases at the domain boundaries in contrast to a comparatively non-polar species like polystyrene.

The mechanical properties of two interstitially modified polyurethane elastomers are shown in Table 6.

TABLE 6
THE   MECHANICAL   PROPERTIES   OF   TWO   INTERSTITIALLY   MODIFIED   POLYURETHANE ELASTOMERS
(24 % Urethane Hard Block, 20 % Hard Polyvinyl Phase)

System	Polyester MDI glycol, 20 % cross-linked methacrylate	Polyester MDI glycol, 20 % cross-linked styrene
Hardness, Shore A °	88	85
Tensile strength, $MN\,m^{-2}$	49	42
Elongation, %	600	715
100 % modulus, $MN\,m^{-2}$	9	6
300 % modulus, $MN\,m^{-2}$	22	11
Angle tear strength, $N\,m^{-1} \times 10^{-3}$	116	108

(c) *The Sequential Formation of the Blend Polymer Followed by the Polyurethane*

This method depends on the formation of the blend polymer in one or more of the polyurethane precursors. To minimise processing problems it is important that the blend polymer does not significantly increase the viscosity of the resins. For this reason the polymer is most usefully present as a stable dispersion.

Most work along these lines has been devoted to the so-called 'polymer polyols' which are dispersions of vinyl polymers in a polyether. These have been known for several years[80,81] and are made by the *in situ* polymerisation of the vinyl monomer in a liquid polyether usually based on propylene oxide.[82,83] Polymer polyols based on acrylonitrile or acrylonitrile copolymers show much greater dispersion stability than those based on most other vinyl polymers and usually 20–30 % of the dispersed phase is incorporated.

Stabilisation of the dispersion is obtained by the formation of some graft copolymer by the abstraction of a hydrogen atom from the polyether backbone:

$$R\cdot + \left(\!\!CH_2\!-\!\underset{\underset{H}{|}}{\overset{\overset{CH_3}{|}}{C}}\!-\!O\!\right)_{\!\!m} \;\to\; RH + \left(\!\!CH_2\!-\!\underset{\cdot}{\overset{\overset{CH_3}{|}}{C}}\!-\!O\!\right)_{\!\!m}$$

$$\Bigg\downarrow \begin{array}{l} n+1 \\ CH_2\!=\!CHCN \end{array}$$

$$\left(\!\!CH_2\!-\!\underset{\underset{(CH_2CHCN)_n}{|}}{\overset{\overset{CH_3}{|}}{C}}\!-\!O\!\right)_{\!\!m}$$
$$\underset{CH_2CH_2CN}{|}$$

This graft copolymer prevents flocculation of the dispersion by the process of steric stabilisation.[84] The formation of this graft copolymer is a random process and the resulting dispersions tend to be coarse (usually about 1 μm in diameter and non-uniform). Much improved dispersions can be made by the controlled addition of a preformed stabiliser and very much finer, uniformly spherical particles can be obtained of about 5000 Å diameter[78] (Fig. 8).

Surprisingly these much finer particles show no advantage in the properties of the final polymer composite. Little reinforcement (except in modulus) of the elastomers is found and the elongation at break is significantly reduced whatever the size of dispersion. This is in contrast to the formation of the polyblends by the interstitial polymerisation method.

Polymer polyols have principally found application in the high-resilience flexible foam area where they are used for their modulus-enhancing properties. This is discussed in greater detail in Chapter 4.

(d) Blends of Thermoplastic Polymers

This method has the considerable merit over the previous methods of

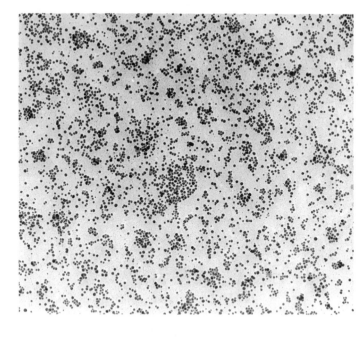

FIG. 8. Polymer polyol dispersions. Left: 20% acrylonitrile-co-styrene in poly(propylene glycol), no added stabilisers (×10 000). Right: 20% acrylonitrile-co-styrene in poly(propylene glycol), using added stabiliser (×10 000).

simplicity, requiring only an extruder or mixer to form the polyblend. It is, however, limited to thermoplastic polymers which have similar processing temperature ranges and have sufficient compatibility to remain as stable blends during a moulding cycle. Despite these limitations several thermoplastic urethane blends are known and a few are of commercial importance.

Blends of polyurethanes with PVC have been known for many years.[85] PVC is very temperature sensitive and care has to be taken with the blending. PVC blends have found particular application for combining the toughness and abrasion resistance of the polyurethane with the stiffness and high modulus of PVC, whilst at the same time cheapening the polyurethane. Plasticised PVC can also be used and offers some advantages in processing.

The use of chlorinated polyethylene as a blend polymer has been disclosed, either alone[86] or in a mixture with PVC[87,88] or with polyethylene,[89] in the latter case the chlorinated polyethylene acting as a carrier for the highly incompatible polyethylene.

Styrene–acrylonitrile copolymers (SAN) and the rubber modified versions, ABS, have also been blended with thermoplastic polyurethanes by co-extrusion.[85] Again the stiffness and initial modulus are improved without much sacrifice of the low temperature properties.

Other polymers which have been blended with polyurethanes include acetal,[85] cellulose propionate,[85] polyether polysulphone resins,[90] bisphenol A-epichlorhydrin resins[91] and poly(tetramethylene terephthalate).[92,93]

Interest has recently been shown in incompatible polymer blends[94] and it is to be expected that, if a stable mixed morphology can be maintained during a thermoplastic moulding cycle, such blends could show interesting properties as well as offering possibly cheaper polyurethanes. A third polymer, copolymer or graft copolymer can be used as a carrier or surfactant[95] to maintain the polymer dispersion during moulding to prevent lamination.

DEVELOPMENTS IN APPLICATION TECHNOLOGY

Solid Reaction Moulded Polyurethanes
Since their commercialisation over twenty years ago, reaction moulded or cast polyurethane elastomers have become well-established and accepted as materials with outstanding wear, abrasion and oil resistance.

Although cold cure and one-shot systems have been developed most of these elastomers are formed by reaction of a liquid prepolymer made from a polyether or polyester and the diisocyanate, with a diol or diamine chain extender. The range of systems now available is very wide, many manufacturers offering a series of prepolymers each of which may be chain extended with several alternative extenders in various proportions to give a range of hardnesses from about 50° Shore A to 70° Shore D. These systems are of three principal types: those based on naphthalene diisocyanate (NDI), those based on TDI and those based on MDI.

The NDI systems are the original 'Vulkollan' elastomers which still retain a significant share of the market due to their excellent load-bearing and heat-resistant characteristics. The prepolymer is based on a polyester but is unstable and has to be made by the fabricator just prior to use. Diols are used as the chain extenders.

The TDI systems may be based on either polyethers or polyesters but in order to obtain adequate phase separation for good elastomeric properties, aromatic diamines are generally used as the extenders to give polyurea hard blocks. The preferred diamine is 3,3'-dichloro-4,4'-diaminodiphenyl-methane (MOCA) which confers an acceptable balance of reactivity and good physical properties. MOCA is now suspected of being a health hazard[96-98] and alternatives are being developed.[99] DuPont have disclosed the use of complexes of 4,4'-diaminodiphenylmethane with selected salts, the complexes dissociating to give the free amine at the cure temperatures.[100-102]

Most of the more recently developed cast elastomer systems are based on MDI. MDI prepolymers are storage stable, in contrast to the NDI types, and also have the advantage of being available from several manufacturers. They also present less of a health hazard than the TDI, MOCA based systems. The MDI systems are based on either polyethers or polyesters usually with diol chain extenders.

In all three types of systems the elastomer may contain some chemical cross-linking. This is introduced either by the use of an excess of the diisocyanate over the hydroxyl (and amine) groups, this excess forming allophanate or biuret linkages by reaction with urethane or urea groups, respectively, or by the incorporation of some polyfunctional chain extender, polyether or polyester.

Mixing of the prepolymer and chain extender is either by batch mixing or continuously mixing in a mixing head usually at a temperature of 80 to 120°C. Fabrication involves pouring the mix into moulds, although other techniques such as rotational and centrifugal casting, spraying and compression moulding may be used.[51] After a short time in the mould (typically 50 to 60 min) the casting, although still relatively weak, is removed from the mould and final curing carried out in an oven.

Solid reaction moulded polyurethane elastomers have found useful application in a variety of areas where good wear properties, high strength and oil and solvent resistance are required. These can be conveniently described under the following headings: rollers; wear parts; tyres; and cold-cure applications.

Rollers

Soft polyurethane rollers (from about 15° Shore A to about 55° Shore A) have been in use for many years mainly in the printing industry. These polyurethanes are generally made by a one-shot technique and are usually based on polyesters chain extended with TDI. They are one-phase amorphous elastomers with a very low compression set ($< 10 \%$) and good ink and solvent resistance.

Recently, harder polyurethane rollers have been developed (from about 55° Shore A to about 95° Shore A) for more demanding mechanical handling applications, for example, in the steel, textile and paper industries. These elastomers are the normal two-phase block copolymers made from either MDI prepolymers and diols or TDI prepolymers and MOCA. Although having a higher compression set (usually $> 30 \%$), these harder elastomers have outstanding wear and abrasion resistance and have a significantly longer life than conventional steel or other rubber rollers.

Wear Parts

This is a comparatively new area with considerable growth potential. Polyurethane elastomers are now proven to engineers but because of their excellent wear resistance much of the testing and proving in particular applications can take a long time.

Many of these applications are in the mining and quarrying industries for the handling of highly abrasive ores and slurries, for example in separation screens and in linings for pipes, pumps, cyclones and impellers, etc. The polyurethane can prolong the life of the metal parts from three to thirty times depending on the hardness of the system used. Both MDI–diol and TDI–MOCA prepolymer systems are used, although the former is technically superior. Polyether based systems (poly(oxytetramethylene glycol)) are preferred where good hydrolytic stability is also required, for example in the handling of aqueous slurries of clays and minerals.[103] The liquid elastomer mix may be applied by spraying on metal parts or, in the case of pipes, by centrifugal casting[104,105] and systems have been developed with controlled viscosities for these applications.

Tyres

Solid tyres for use on slow-moving industrial vehicles such as fork-lift trucks are a traditional outlet for reaction moulded polyurethane elastomers. The market is dominated by the polyester–NDI–diol systems because of their superior load-bearing properties, although some TDI–MOCA systems are used. However, the market for solid tyres for slow-moving vehicles is a very small part of the tyre business and much interest was created in 1975 by the announcement by Polyair in Austria of the development of a polyurethane car tyre.[106] The tyre is made by liquid injection moulding technology and is a two-stage moulded unit with the carcase moulded first and then, in a sequential operation, a tread belt is moulded directly onto the carcase.[107] No publication in either the general literature or in the patent literature has yet appeared giving any indication of the polymer chemistry involved. The tyres are claimed to have excellent wear characteristics and 'run flat' qualities as well as being significantly lower in production costs than conventional tyres. It is too early to know whether the polyurethane car tyre can eventually be technically and commercially successful, but it will clearly be some time before it can compete with conventional natural rubber radials.

Cold Cure Applications

Cold cure polyurethane elastomers are not new but several interesting developments have been taking place in the last few years. Most systems use a one-shot technique and are based on MDI and either polyethers or polyesters. Depending on the application, additives such as solid fillers or plasticisers may also be incorporated. Some recent developments have been in the use of crude or polymeric MDI and other liquid variants of MDI which are now finding application in microcellular polyurethane elastomers (see the next section).

Cold cure elastomers are used for cable jointing and potting compounds, for sealants, commercial vehicle mats, coatings for oil discharge hoses and, more recently, for moulds for precast concrete and for tyre filling.

These tyre filling compositions replace the air in pneumatic tyres by a soft, resilient, non-cellular rubber so that tyre deflation and under-inflation are prevented. They are particularly useful for vehicles operating under heavy off-the-road conditions in quarries, docks, lumber yards, etc., where sustained speeds over about 35 mph are not usual. As well as reducing vehicle down time caused by flat tyres, the replacement of air by rubber in the tyres gives a lower centre of gravity on the vehicle and improves stability. Systems have been developed with hardnesses in the range 30 to 40° Shore A. The liquid ingredients are injected through the valve, the tyre being vented at the highest point with a needle. The tyres are ready for use within one to two days after filling.

Microcellular Reaction Moulded Polyurethanes

Most of the dramatic increase in consumption of polyurethane elastomers since 1970 has been due to the advent of reaction moulded microcellular systems having densities from about 0·5 to 0·8 g/cc.

The high pressure processing types (liquid injection moulded, LIM, or reaction injection moulded, RIM) used in flexible automobile parts account for an increasing proportion of this total and are more fully discussed in Chapter 5. The slightly slower curing types which rely on low pressure metering and mixing machinery are mainly used in the manufacture of polyurethane shoe soles.

World consumption, in 1976, of microcellular polyurethanes for shoe soling is estimated at 80 000 tonnes.[108] However, high though this figure is, it should be borne in mind that it represents only 2–4 % of the total shoe-soling market. The market for shoe soling is extremely large and dominated by traditional materials, such as leather, natural rubber, wood and cork as well as the cheap synthetic polymers such as PVC and resin rubbers. Microcellular polyurethanes are able to compete with these materials because of their good abrasion resistance and flexing properties at low densities.

Microcellular polyurethanes for shoe soling are of both the polyester and polyether types. Polyester types were developed first and owing to their generally superior mechanical properties they tend to be used in the thinner sole, higher quality section of the market, although several very low density stiff systems are used, mainly in Italy, for fashion shoes.

Polyester systems are usually based on quasi-prepolymers (made from part of the polyester and the MDI) and a resin blend made up of the rest of the polyester, the chain extender (usually butane diol), catalysts, surfactants and water to produce carbon dioxide to blow the foams. The use of a quasi-prepolymer enables the two component streams to be of similar volume and, perhaps more importantly, of similar viscosity, thereby simplifying and improving the mixing and metering of the streams. Most polyesters are solid or semi-solid at room temperature and heated lines and tanks are therefore necessary on the processing machines. The mixed streams are generally allowed to free-fall from a traversing head into the mould which is then rapidly closed. Foaming and curing take about two to three minutes. On some machines the mixed streams may be injected into the mould under positive pressure, often with the shoe upper held in the mould, thus giving a 'moulded on' sole.

To enable uninterrupted production of unit soles a carousel of twenty to thirty moulds is usually used, curing, demoulding and respraying the mould with release agent being complete by the time the original mould comes round again for refilling.

The principal problem with the polyester systems has been the need for melting out the resins from the drum and the need for heated lines and tanks on processing machines. Some of the earlier systems also suffered from consistency problems and inadequate flex lives of the derived elastomers, particularly at low temperatures. These problems have now been generally overcome and the key parameters affecting the curing, storage stability of the resins and the physical properties of the shoe sole units are now well understood.[109,110] Trends in polyester systems are now towards easier processing types. These are generally based on liquid or low melting polyesters which require much shorter melting out times and remain liquid for longer periods.

As a result of the early processing problems with the polyester systems, polyether based systems were introduced in 1972. These are based on low viscosity liquid polyethers and a liquid MDI prepolymer. Early ether systems gave easier material handling and machine processing than equivalent polyester types but led to units with lower mechanical properties. Many improvements have been made, however, without the loss of the processing advantages and now the world-wide split between polyester and polyether systems is probably about 50:50.

The polyether systems generally consist of a liquid MDI prepolymer and a resin blend containing the polyether, chain extender, catalysts and a fluorcarbon blowing agent. The two streams are mixed in approximately a 1:2 volume ratio.

In contrast to the water blown polyester systems, these fluorcarbon blown systems give integral skin units which have a fairly thick skin formed by the condensation of the fluorcarbon at the mould surface and a much lower density core which at the centre may be 0·4 g/cc or less. Water blown types do not have a skin as such and have a much smaller density change across the units from face to face.

The liquid MDI prepolymer is formed by the reaction of a glycol[111-113] or a mixture of glycols[114,115] with an excess of MDI and remains liquid for several months at room temperature.

The polyethers generally have a functionality greater than two to introduce some chemical cross-linking into the systems to improve the set-up and stiffness at demould. For fast reaction with the isocyanate the polyethers are usually poly(ethylene oxide) tipped to give a high proportion of primary hydroxyl groups on the poly(propylene oxide) based backbone.

The polyether based systems tend to be inherently lower in mechanical properties than the polyester types and this is particularly noticeable at demould when the units can still be soft and deformable. The principal problem with these systems now is the long-term stability of the liquid isocyanate prepolymers and there is a move towards more local manufacture and blending.

The viscosity of the prepolymer tends to increase slowly with time and is in any case higher than the polyether resin blend resulting in mixing problems. Lower viscosity isocyanates similar to those used in LIM and RIM technology may be developed to overcome some of these problems. These isocyanates are made by the partial conversion of MDI to carbodiimides using a phosphorus based catalyst.[116,117] MDI adds to the carbodiimides to form uretonimines[118] which are particularly effective at disrupting the crystallinity of the MDI leading to very low viscosity products with a functionality greater than two:

$$2\,R\text{---}NCO \quad \xrightarrow[\text{(part reaction)}]{\substack{\text{phosphorous} \\ \text{catalyst}}} \quad R\text{---}N\text{=}C\text{=}N\text{---}R + CO_2$$
(MDI)
plus some low polymers
plus free MDI

$$R\text{---}N\text{=}C\text{=}NR + RNCO \rightleftharpoons R\text{---}N\text{=}C\text{---}N\text{---}R$$
(MDI)

R—N—C=O

trifunctional in NCO
plus free MDI

$$R\text{=}OCN\text{---}\bigcirc\text{---}CH_2\text{---}\bigcirc\text{---}$$

These low viscosity isocyanates can be used either alone or in a mixture with glycol modified MDI to give a range of products of varying viscosity and functionality.

Thermoplastic Polyurethanes

Since their commercialisation over twenty years ago, thermoplastic polyurethanes have shown a steady if unspectacular growth. Present West European consumption is estimated at 11 000–12 000 tonnes per annum. Thermoplastic polyurethanes (TPUs) are of both the polyether and polyester types[119] although polyester types account for a larger proportion of the market. Polycaprolactone based systems are becoming of increasing importance[120] in view of their good balance of mechanical properties, hydrolytic stability and cost.

Thermoplastic polyurethanes are made by metering and mixing the ingredients using either a prepolymer or one-shot technique and casting onto trays or onto a moving belt. After complete or partial curing the polymer is granulated. As much of the curing reaction occurs in a solid matrix, local inhomogeneities can occur and some chemical cross-links may be found by allophanate formation. The polymer may therefore be subsequently extruded and sold as lace, or die face cut, granules.

Thermoplastic polyurethanes are more elegantly prepared using a twin-screw extruder. The metered reactants are fed into the extruder and the reactions carried out at such temperatures as to maintain the forming polymer in the melt. Extruded polymer is obtained directly.

In either process additives are incorporated, usually in the precursors or in a subsequent blending operation. These include antioxidants,[121] UV stabilisers,[122] anti-hydrolysis agents[123] (usually polymeric carbodiimides), lubricants, anti-blocking agents, pigments, fillers and plasticisers.

Thermoplastic polyurethanes are now available to cover all thermoplastic processing techniques (injection moulding, blow moulding, extrusion, calendering, etc.), by control of the urethane reactions during manufacture and choice of reactants and additives.

At present the largest single outlet for thermoplastic polyurethanes in Western Europe is in injection moulded ski boots, this now accounting for nearly 60% of TPU production. Thermoplastic polyurethanes have established themselves in this market owing to their excellent abrasion resistance and low temperature flexing and impact properties, although their cost has limited penetration to the quality and competition part of the market. Generally very hard systems are used (55° Shore D and above) and they are usually based on polyesters.

Thermoplastic polyurethanes also find widespread application in the general automotive area for items such as seals and grommets due to the excellent abrasion and oil resistant properties of the elastomers, although 'under the bonnet' applications are limited by the upper service temperature limit of 70–80 °C for these polymers. One particular application of increasing importance is in blow moulded protection bellows and gaiters for exposed steering joints.

A potentially very large market exists for thermoplastic polyurethanes in competition with reaction moulded types for soft exterior parts for automobiles and polymers are being designed for this application.[124]

Extrusion grades of polyurethane have found widespread application for cable coverings and hydraulic hoses and sheathings, again due to their toughness, abrasion and oil resistance and good low temperature flex properties. Polyurethane sheathing is used, for example, in outside broadcast TV cables, umbilicals and blow-out preventer hoses.

Extrusion grades are available for the manufacture of polyurethane film by extrusion through a slit die. Polyurethane film can also be made by a 'lay flat blown bubble' process or by a calendering process (see the next section). Polyurethane film finds application in packaging, fabric coating, belting, etc.

There is increasing interest in medical grades of thermoplastic polyurethanes and these are usually based on poly(oxytetramethylene) glycol soft blocks. The market is as yet small but has considerable potential for a wide range of disposable items and polyurethanes could partly replace PVC in several applications where plasticiser migration, or low temperature properties, present problems.

More speculatively, microcellular grades of TPU have been produced although none are available commercially. The main problem with injection moulded microcellular types is that of excessively long moulding cycles but, if this can be overcome, the market could be very large. Microcellular products have been made by extrusion processes.

Coatings

The use of elastomeric polyurethanes for coating leather and fabrics is now well established. In 1975 the US consumption of polymer for leather finishes was estimated at 2200 tonnes and for fabric coating at 11 000 tonnes.[125]

The use of polyurethanes for fabric coating is increasing rapidly and this area has been the subject of several recent developments.

At present, most polyurethane fabric coatings are applied from solution either as two-pack reactive types or one fully cured type. The fabric may be coated either directly or, more usually, by prior coating onto release paper and subsequent transfer of the film onto the fabric. Direct coatings are generally cheaper to apply but they produce composites with stiffer 'hand' and the range of fabric substrates is limited. Transfer coating provides softer composites which are easily embossed and simulate a leather look but the costs are higher.

With the increasing environmental pressures and the high cost of solvents, several non-solvent or 100 % solids processes are being developed. These fall into three types of process; direct or transfer coating using 100 % solids reactive systems; calendering of thermoplastic urethanes and film lamination techniques.

100 % Solids Reactive Systems

These systems are based on a two-stream metering and mixing followed by immediate application and spreading on the fabric or release paper. The control of the viscosity and curing rate by suitable choice of catalysts, curing temperatures and reactants is crucial in order to ensure a defect-free coating with a pleasant feel and handle. If the viscosity is too low, the coating penetrates the fabric giving a stiff composite. On the other hand, if gelation is too rapid, streaking, pin-holing and orange peel effects can occur.[126] A further problem with systems of this type is the need for coaters to invest in a new metering and mixing equipment which can involve large additional costs and may offset the potential savings of not using a solvent.

An alternative approach is a two-pot 100 % solids system which can be blended under agitation and have a sufficiently long pot life (8–16 h) so that it can be coated using conventional knife coating techniques. Curing at 150 °C for 2 min gives a tack free coating.[126]

A new system which is similar to this is a 100 % solids, heat curable foam which can be direct or transfer coated with the standard knife or reverse roll coaters. By incorporation of a blowing agent, the urethane coating is foamed during the oven curing. It is claimed to eliminate the need for the use of a napped and sheared fabric and give an excellent 'hand'.[127] It is normally used with the conventional type of top coat.

Hot Melt Calendering

The hot melt process is already being used commercially and is expected to grow in importance during the next decade. Zimmer Plastics and Von Roll make hot melt calenders suitable for use with polyurethanes.[128] These machines differ from conventional PVC calenders in several respects.[129] Owing to the high processing temperatures of thermoplastic polyurethanes, the roll temperatures are high (about 200 °C) and, therefore, to minimise polymer degradation, the residence time of the molten polymer on the rolls is short and the material is subjected to a high friction on the rolls rather than to a low frictional heat build up. Although differing in detail, both

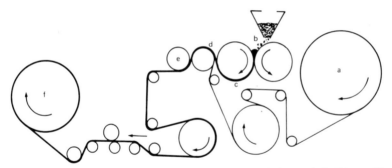

FIG. 9. Diagram of Zimmer melt calender. (a) = Fabric unwind; (b) = rolling bank of molten polymer; (c) = film of molten polymer; (d) = transfer to fabric; (e) = finishing roll, (f) = fabric rewind. Taken from Zimmer Plastic GmbH trade literature.

machines are essentially similar and a diagram of a Zimmer machine is shown in Fig. 9.

The polymer may be dispensed as granules into the nip of the melt rolls or, preferably, as an extended preplasticised 'rope' to increase the throughput rate by reducing the work of the melt rolls.[130] The molten film is transferred to the fabric at a second nip between the melt roll and a rubber take-off roller. The laminated fabric passes under a finishing roll to impart a smooth, matt or embossed finish.

Direct or transfer coating is possible with film thickness down to about 50–70 μm. The composites are usually, therefore, stiffer and of heavier gauge and mainly find application in heavy duty and industrial outlets.

Special grades of thermoplastic polyurethane have been developed for these melt coating machines. These polymers are of lower melt viscosity than normal TPUs and have a wider processing temperature range.[40] Lubricants and release agents are incorporated to give a clean release of the molten polymer from the hot rolls.

Coated fabrics can also be made on conventional calenders in conjunction with a slit die extender, the calender in this case spreading and laminating the molten film.

Film Lamination
This method, which is a logical extension of the above, provides coated fabrics with a good 'feel' since the polyurethane does not penetrate between the weave of the fabric.

The films may be made by extrusion or calendering and a variety of methods are available for bonding the film to the fabric, the relatively high cost of the film being offset by the low conversion costs. Solvent and latex adhesives may be used in conjunction with oven or drum laminators and the composites can be easily embossed and printed.[131]

Flame lamination can also be used to bond urethane films to urethane foam. The surface of the foam is partially degraded by the flame and the sticky decomposed polymer is used as an adhesive to bond the film to the foam.[131]

An alternative approach to the problem of the use of organic solvents with urethane coatings is to use aqueous emulsions of polyurethanes. This technique has been the subject of considerable development work particularly by Bayer workers.[132]

The polyurethanes contain ionisable groups in the hard blocks capable of salt formation. Groups such as quaternary ammonium, phosphonium, tertiary sulphonium[133,134] or sulphonate[135] and carboxylate[136] have been used. These ionic polyurethanes, while not being salt-like enough to be soluble in water, are insoluble in most common low boiling organic solvents. They are, however, soluble in solvents such as acetone, containing some water. Further water may be added and the acetone removed to give stable aqueous polyurethane latices with particle sizes from about 5 nm to about 10 μm.[132]

The absence of any emulsifiers and surfactants results in excellent film formation with surfaces with a high gloss. These polyurethane latices are mainly finding application as adhesive tie coats—for transfer coated fabrics—but many more applications will undoubtedly be developed as restrictions on organic solvents increases. Various other methods for the preparation of aqueous polyurethane emulsions have been described.[137]

A further non-solvent method for coating with polyurethanes is the use of polyurethane powders. As yet this has little commercial use but is expected to grow in importance in the next decade. Powders may be of the reactive type using blocked isocyanates[138] or be powdered thermoplastic polymers prepared either by cryogenic grinding or by direct emulsion polymerisation methods in inert organic liquids.[139-42]

Polyurethane powders have been used as melt adhesives and could find extensive use for coating metal articles by fluid bed, spray coating or electrostatic techniques.[143]

REFERENCES

1. KALLERT, W. (1976). Chemistry in Industry—The Way Ahead, Conf., Wembley, UK. Nov.

2. BONART, R. (1968). *J. Macromol. Sci., Phys.*, **2**, 115.
3. CLOUGH, S. B., SCHNEIDER, N. S. and KING, A. O. (1968). *J. Macromol. Sci., Phys.*, **2**, 64.
4. OGURA, K. and SOBUE, M. (1972). *Polym. J.*, **3**, 153.
5. ANDREAS, R. D. and HAMMACK, T. J. (1965). *J. Polym. Sci.*, B, **3**, 665.
6. TOBOLOSKY, A. V. and SHEN, M. C. (1963). *J. Phys. Chem.*, **67**, 1886.
7. OTOCKA, E. P. and EIRICH, F. R. (1968). *J. Polym. Sci.*, A-2, **6**, 895.
8. FITZGERALD, W. E. and NIELSEN, L. E. (1964). *Proc. Roy. Soc.*, **A282**, 137.
9. SEYMOUR, R. W., ESTES, G. M. and COOPER, S. L. (1970). *Macromol.*, **3**, 579.
10. PAIK SUNG, C. S. and SCHNEIDER, N. S. (1975). *Macromol.*, **8**, 68.
11. ISHIHARA, M., KIMURA, I., SAITO, K. and ONO, M. (1974). *J. Macromol. Sci., Phys.*, **10**, 591.
12. ALLPORT, D. C. and MOHAJER, A. A. (1973). In *Block Copolymers*, D. C. Allport and W. H. Janes, eds, Applied Science Publishers Ltd, London, Chapter 8C.
13. CLOUGH, S. B. and SCHNEIDER, N. S. (1968). *J. Macromol. Sci., Phys.*, **2**, 553.
14. MILLER, G. W. and SAUNDERS, J. H. (1970). *J. Polym. Sci.*, A-1, **8**, 1923.
15. VROUENRAETS, C. M. F. (1972). *Polym. Prepr.*, **13**, 529.
16. SEYMOUR, R. W. and COOPER, S. L. (1973). *Macromol.*, **6**, 48.
17. MACKNIGHT, W. J., YANG, M. and KAJIYAMA, T. (1968). *Polym. Prepr.*, **9**, 860.
18. BONART, R., MORBITZER, L. and MULLER, E. H. (1974). *J. Macromol. Sci., Phys.*, **9**, 447.
19. HOYER, H. Unpublished work cited in Ref. 16.
20. SCHNEIDER, N. S., DESPER, C. R., ILLINGER, J. L. and KING, A. O. (1975). *J. Macromol. Sci., Phys.*, **11**, 527.
21. SAMUELS, S. L. and WILKES, G. L. (1973). *J. Polym. Sci. Polym. Symp.*, **43**, 149.
22. SEYMOUR, R. W., ALLEGREZZA, A. E. and COOPER, S. L. (1973). *Macromol.* **6**, 896.
23. KIMURA, I., ISHIHARA, H., ONO, H., YOSHIHARA, N., NOMURA, S. and KAWAI, H. (1974). *Macromol.*, **7**, 355.
24. COOPER, S. L., WEST, J. C. and SEYMOUR, R. W. (1976). *Encyclopedia of Polymer Science and Technology*, Supplement vol. 1.
25. WILDNAUER, R. and WILKES, G. L. (1975). *Polym. Prepr.*, **16**, 600.
26. WILDNAUER, R. and WILKES, G. L. (1975). *J. Appl. Phys.*, **46**, 4148.
27. EMERSON, J. A. and WILKES, G. L. (1976). *J. Appl. Phys.*, **47**, 4261.
28. TRELOAR, L. R. G. (1958). *Physics of Rubber Elasticity*, 2nd edn., Oxford University Press, Oxford.
29. JEFFS, G. M. F. J. Imperial Chemical Industries Ltd, unpublished work.
30. TRAPPE, G. (1968). In *Advances in Polyurethane Technology*, J. M. Buist and M. Gudgeon, eds, Maclaren, London.
31. KAJIYAMA, T. and MACKNIGHT, W. J. (1968). *Trans. Soc. Rheol.*, **13**, 527.
32. SEEFRIED, C. G., KOLESKE, J. V. and CRITCHFIELD, F. E. (1975). *J. Appl. Polym. Sci.*, **19**, 2493.
33. SEEFRIED, C. G., KOLESKE, J. V. and CRITCHFIELD, F. E. (1975). *J. Appl. Polym. Sci.*, **19**, 2503.
34. SEEFRIED, C. G., KOLESKE, J. V. and CRITCHFIELD, F. E. (1975). *J. Appl. Polym. Sci.*, **19**, 3185.

35. SEEFRIED, C. G., KOLESKE, J. V., CRITCHFIELD, F. E. and DODD, J. L. (1975). *Polym. Eng. and Sci.*, **15**, 646.
36. ILLINGER, J. L. and SCHNEIDER, N. S. (1972). *Polym. Eng. and Sci.*, **12**, 25.
37. AITKEN, R. R. and JEFFS, G. M. F. (1977). *Polymer*, **18**, 197.
38. JEFFS, G. M. F. and REDMAN, R. P. Imperial Chemical Industries Ltd, unpublished work.
39. BROWN, J. P., JEFFS, G. M. F. and REDMAN, R. P. Imperial Chemical Industries Ltd, unpublished work.
40. FARBENFABRIKEN BAYER, British Patent 1,421,808 (3.10.73).
41. METZGER, S. M. and PREPELKA, D. J. (1976). *J. Elast. Plast.*, **8**, 141.
42. EVANS, M. C., CRITCHFIELD, F. E. and GERKIN, R. M. (1976). *J. Cell. Plast.*, **12**, 235.
43. UNION CARBIDE CORP., British Patent 1,233,614 (25.6.68).
44. BONART, R., MORBITZER, L. and RINKE, M. (1970). *Kolloid Z.*, **240**, 807.
45. GIANATASIO, P. A. and FERRARI, R. J. (1966). *Rubber Age*, **98**, 83.
46. BEUCHE, F. (1965). In *Reinforcement of Elastomers*, G. Krause, ed., InterScience, New York, Chapter 2.
47. UNION CARBIDE CORP., British Patent 1,388,748 (2.8.71).
48. UNION CARBIDE CORP., British Patent 1,389,039 (2.8.71).
49. BRIDGESTONE TIRE CO. LTD, German Patent 2,261,482 (15.12.72).
50. BRIDGESTONE TIRE CO. LTD, German Patent 2,432,029 (3.7.73).
51. WRIGHT, P. and CUMMING, A. P. C. (1969). *Solid Polyurethane Elastomers*, Maclaren, London.
52. TILLEY, P. N., NADEAU, M. G., REYMORE, H. E., WASZECIAK, P. H. and SAYIGH, A. A. R. (1968). *J. Cell. Plast.*, **4**, 56.
53. SCHOLLENBERGER, G. S. and DINBERGS, K. (1974). *Adv. Urethane Sci. & Technol.*, **3**, 36.
54. MOBAY CHEM. CO., US Patent 3,310,533 (2.1.62).
55. FARBENFABRIKEN BAYER, German Patent 2,418,075 (13.4.74).
56. FARBENFABRIKEN BAYER, British Patent 1,438,145 (12.5.72).
57. MANSON, S. A. and SPERLING, L. H. (1976). *Polymer Blends and Composites*, Heyden, New York.
58. KLEMPNER, D. and FRISCH, K. C. (1974). *Adv. Urethane Sci. & Technol.*, **3**, 14.
59. CORISH, P. J. and POWELL, B. D. W. (1974). *Rubber Chem. and Technol.*, **47**, 481.
60. FRISCH, K. C., KLEMPNER, D., ANTCZAK, T. and FRISCH, H. L. (1974). *J. Appl. Polym. Sci.*, **18**, 683.
61. FRISCH, K. C., KLEMPNER, D., MIGDAL, S., FRISCH, H. L. and DUNLOP, A. P. (1975). *J. Appl. Polym. Sci.*, **19**, 1893.
62. FRISCH, K. C., KLEMPNER, D., MIDGAL, S. and FRISCH, H. L. (1974). *J. Polym. Sci.*, A-1, **12**, 885.
63. FRISCH, H. L., FRISCH, K. C. and KLEMPNER, D. (1974). *Polym. Eng. and Sci.* **14**, 646.
64. FRISCH, K. C., KLEMPNER, D., MUKHERJEE, S. K. and FRISCH, H. L. (1974). *J. Appl. Polym. Sci.*, **18**, 689.
65. KIM, S. C., KLEMPNER, D., FRISCH, K. C., FRISCH, H. L. and GHIRADELLA, H. (1975). *Polym. Eng. and Sci.*, **15**, 339.
66. KIM, S. C., KLEMPNER, D., FRISCH, K. C., RADIGAN, W. and FRISCH, H. L. (1966). *Macromol.*, **9**, 258.

67. KIM, S. C., KLEMPNER, D., FRISCH, K. C. and FRISCH, H. L. (1976). *Macromol.*, **9**, 263.
68. FRISCH, H. L., KLEMPNER, D. and FRISCH, K. C., (1969). *J. Polym. Sci.*, B, **7**, 775.
69. KLEMPNER, D., FRISCH, H. L. and FRISCH, K. C. (1970). *J. Polym. Sci.*, A-2, **8**, 921.
70. MATSUO, M., KWEI, T. K., KLEMPNER, D. and FRISCH, H. L. (1970). *Polym. Eng. and Sci.*, **10**, 327.
71. KLEMPNER, D. and FRISCH, H. L. (1970). *J. Polym. Sci.*, B, **8**, 525.
72. KAPLAN, D. S. (1976). *J. Appl. Polym. Sci.*, **20**, 2615.
73. ALLEN, G., BOWDEN, M. J., LEWIS, G., BLUNDELL, D. J. and JEFFS, G. M. (1974). *Polymer*, **15**, 13.
74. ALLEN, G., BOWDEN, M. J., LEWIS, G., BLUNDELL, D. J., JEFFS, G. M. and VYVODA, J. (1974). *Polymer*, **15**, 19.
75. ALLEN, G., BOWDEN, M. J., TODD, S. M., BLUNDELL, D. J., JEFFS, G. M. and DAVIES, W. E. A. (1974). *Polymer*, **15**, 28.
76. ALLEN, G., BOWDEN, M. J., BLUNDELL, D. J., HUTCHINSON, F. G., JEFFS, G. M. and VYVODA, J. (1973). *Polymer*, **14**, 597.
77. ALLEN, G., BOWDEN, M. J., BLUNDELL, D. J., JEFFS, G. M., VYVODA, J. and WHITE, T. (1973). *Polymer*, **14**, 604.
78. REDMAN, R. P. Imperial Chemical Industries Ltd, unpublished work.
79. IMPERIAL CHEMICAL INDUSTRIES LTD, British Patent 1,440,068 (30.10.72).
80. FARBWERKE HOECHST, British Patent 922,457 (15.4.58).
81. FARBWERKE HOECHST, British Patent 969,965 (5.1.60).
82. KURYLA, W. C., CRITCHFIELD, F. E., PLATT, L. W. and STAMBERGER, P. (1966). *J. Cell. Plast.*, **2**, 84.
83. CRITCHFIELD, F. E., KOLESKE, J. V. and PRIEST, D. C. (1972). *Rubber Chem. and Technol.*, **45**, 1467.
84. OSMOND, D. W. J. and WAITE, F. A. (1975). *Dispersion Polymerisation in Organic Media*, K. E. J. Barrett, ed., Wiley, London, Chapter 2.
85. BONK, M. W., SARDANOPOLI, A. A., ULRICH, H. and SAYIGH, A. A. R. (1971). *J. Elastoplast.*, **3**, 157.
86. UNIROYAL INC., Canadian Patent 980,936 (19.3.73).
87. UNIROYAL INC., Canadian Patent 980,487 (19.3.73).
88. UNIROYAL INC., US Patent 3,882,191 (29.3.73).
89. UNIROYAL Inc., US Patent 3,929,928 (29.3.73).
90. UNIROYAL INC., British Patent 1,436,014 (31.8.72).
91. B. F. GOODRICH CO., German Patent 1,694,269 (29.4.65).
92. DAINIPPON INK CHEM. KK., Japanese Patent 0053,448 (13.9.73).
93. TOYOBO KK., Japanese Patent 0079,559 (16.11.73).
94. SEEFRIED, C. G., KOLESKE, J. V. and CRITCHFIELD, F. E. (1976). *Polym. Eng. and Sci.*, **16**, 771.
95. BASF AG, British Patent 1,440,030 (19.9.72).
96. *Rubber Journal*, 1969, **151**, 16.
97. *Rubber Age*, 1971, **103**, 78.
98. *Plastics and Rubber Weekly*, 1971, 30th July, 1; 13th August, 6.
99. BOLGER, J. C. and CHILDS, W. L. (1974). *Soc. Plast. Eng. Tech. Pap.*, **20**, 211.
100. DUPONT DE NEMOURS & CO., British Patent 1,400,921 (1.5.72).
101. DUPONT DE NEMOURS & CO., British Patent 1,435,845 (14.2.73).

102. Du Pont de Nemours & Co., British Patent 1,462,418 (1.6.73).
103. Munro, C. E. (1974). *Pipes, Pipelines Int.*, **19**, 16.
104. English Clays, Lovering Pochin & Co., British Patent 1,444,225 (14.9.72).
105. English Clays, Lovering Pochin & Co., British Patent 1,444,908 (14.9.72).
106. *Plastics and Rubber Weekly*, 1975, 14th March, 64.
107. *European Rubber Journal*, 1976, **158**, 10.
108. English, G. Imperial Chemical Industries Ltd, Private Communication.
109. Carleton, P. S., Ewen, J. M., Reymore, H. E. and Sayigh, A. A. R. (1974). *J. Cell. Plast.*, **10**, 283.
110. Carleton, P. S. (1975). *Soc. Plast. Eng. Tech. Pap.*, **21**, 555.
111. Farbenfabriken Bayer, British Patent 1,158,534 (8.3.67).
112. Upjohn, British Patent 1,087,388 (24.10.65).
113. Upjohn, British Patent 1,087,389 (24.10.65).
114. Imperial Chemical Industries Ltd, British Patent 1,378,975 (26.9.72).
115. Imperial Chemical Industries Ltd, British Patent 1,377,676 (21.8.72).
116. Arnold, R. G., Nelson, S. A. and Verbanc, J. J. (1957). *Chem. Revs.*, **57**, 47.
117. Imperial Chemical Industries Ltd, British Patent Applications 13656/75 (3.4.75) 13657/75 (3.4.75), 13658/75 (3.4.75), 13659/75 (3.4.75), 15649/75 (16.4.75) and 44965/75 (30.10.75).
118. Farbenfabriken Bayer, British Patent 795,720 (24.6.55).
119. Gee, H. L. (1975). *J. Coated Fab.*, **4**, 191.
120. Critchfield, F. E., Koleske, J. V. and Dunleavy, R. A. (1971). *Rubber World*, **164**, 61.
121. Fabris, H. J. (1976). *Adv. Urethane Sci. & Technol.*, **4**, 89.
122. Schollenberger, C. S. and Stewart, F. D. (1976). *Adv. Urethane Sci. & Technol.*, **4**, 68.
123. Farbenfabriken Bayer, British Patent 986,200 (2.12.60).
124. Hoodbhoy, A. I. (1974). *J. Elast. Plast.*, **6**, 269.
125. Bedoit, W. C. (1974). *J. Cell. Plast.*, **10**, 78.
126. Krishnan, S. (1976). *J. Coated Fabrics*, **6**, 39.
127. Scott, P. H. and Carey, D. A., *J. Coated Fabrics*, **6**, 13.
128. Mann, A. (1975). *J. Coated Fabrics*, **5**, 133.
129. *Modern Plastics Int.*, 1971, Sept., 18.
130. Adank, G. (1974). *Polymer Age*, **5**, 15.
131. Pittenger, F. (1976). *J. Coated Fabrics*, **6**, 85.
132. Dieterich, D., Keberle, W. and Wuest, R. (1970). *J. Oil Col. Chem. Assoc.*, **53**, 363.
133. Farbenfabriken Bayer, British Patent 1,078,202 (19.9.63).
134. Farbenfabriken Bayer, British Patent 1,080,590 (28.12.64).
135. Farbenfabriken Bayer, British Patent 1,462,597 (28.9.74).
136. Farbenfabriken Bayer, British Patent 1,193,732 (2.3.67).
137. New, E. A., Regos, N. and Labb, P. (1976). *J. Elast. Plast.*, **8**, 210.
138. Obendorf (1974). *Paint Manuf.*, **44**, 11.
139. USM Corp., British Patent 1,440,131 (7.8.72).
140. USM Corp., British Patent 1,440,132 (16.6.72).
141. Imperial Chemical Industries Ltd, German Patent 2,215,732 (30.3.71).
142. Union Carbide Corp., US Patent 4,000,218 (17.7.75).
143. Lehmann, W., G.D.R. 113,488 (23.9.74).

Chapter 4

DEVELOPMENTS IN THE USE OF FLEXIBLE URETHANE FOAMS

G. WOODS

ICI Organics Division, Manchester, UK

SUMMARY

Recent developments in the chemistry of the polyols, isocyanates and additives used in flexible foam production are surveyed and their effects on the properties of the foam are indicated.

The performance testing of polyurethane foam is described with emphasis on foam for automotive seating, the trend towards the use of 'cold-cure' moulding systems and the development of mechanical seating.

The hazards involved both in the manufacture and the use of flexible foam are described and methods of control are reviewed.

POLYOLS

Polyester Polyols
The first flexible polyurethane foams were based upon lightly branched polyesters reacted with TDI 65 (65/35 ratio of the 2:4/2:6 isomers) and although these gained some acceptance for upholstery in Europe, the rapid expansion of flexible foam production awaited the introduction of the more resilient polyether polyol foams. The use of polyester foams is now confined largely to applications in clothing, textile lamination, footwear, luggage and vehicle trim where solvent resistance and high load-bearing properties are required. Polyester foam production now represents less than 10 % of all flexible foam slabstock production and is based almost entirely upon poly(diethylene glycol adipates). Recent developments have been aimed at

77

obtaining the optimum combination of reactivity, molecular weight and viscosity to yield maximum conversion rates, expecially in the modern continuous cylindrical slabstock processes (Fig. 1). Typical products are lightly branched with trimethylolpropane or, increasingly, with pentaerithrytol.

FIG. 1.   Cylindrical block polyester foam. (Caligen Foam Ltd, Accrington, UK.)

Polyester polyols for foam-making are made by a number of manufacturers but the polyols available fall into three basic types: polyesters for general purpose foams for use with TDI 65 or TDI 80, polyesters for 'textile' foams with TDI 80, and branched relatively low molecular weight polyesters for making the hard semi-rigid slabstock foam used for cleaning pads and for sun-visors and similar trim fabrications in vehicles. Typical polyester properties are tabulated in Table 1. The general purpose polyesters of molecular weight about 2000 give foams with typical elongation at break in the range from 150% to 300% depending on the density and formulation while textile foam esters, usually made using TDI 80 at an index between 90 and 98 (i.e. with 10% to 2% less than the stoichiometric TDI requirement), yield softer foams with an elongation at break in the range from 350% to 450%. Usually semi-rigid slabstock foam is made using about 50 parts by weight of a general purpose polyester and

TABLE 1
TYPICAL PROPERTIES OF POLYESTERS FOR FLEXIBLE FOAM

Property	General purpose polyester	Round-block and textile foam polyester	Semi-rigid foam polyester
Specific gravity at 25 °C	1·19	1·19	1·18
Hydroxyl value	65	55	230 to 270
Acid value (mg KOH/g)	3	· 1·5	2 to 5
Water content (%)	0·1	0·1	0·1 to 0·2
Viscosity (poise)			
at 5 °C	900 to 1 400	1 600	—
at 15 °C	350 to 400	450	420 to 500
at 25 °C	150 to 180	200	180 to 235
at 40 °C	50 to 60	65	55 to 70
Fire point (°C) (ASTM D92-56)	338	> 315	338
Flash point (°C) (ASTM D92-56)	300	260	300
Typical foam properties	(can be adjusted within the range indicated)		
Density (kg/m³)	21 to 30	21 to 32	24 to 32
Elongation at break (%)	150 to 350	300 to 500	80 to 110
Compression set (%)			
(BS 4443 Method 6A			
at 75% compression)	10 to 5	5 to 12	ca 50
Compression hardness (g/cm²)			
(at 40% comp)	47 to 67	40 to 51	200 to 500
Indentation hardness (N)			
(BS 44431: Part 2)	225 to 325	197 to 245	n.a.

50 parts by weight of a highly branched polyester reacted with TDI 65. The use of TDI 65 is essential to obtain the stiff cross-linked foams without excessive closed cell content which causes shrinkage and distortion of the foam after cutting.

Polyether Polyols
Virtually all upholstery foams are now based upon polyether polyols produced by the base catalysed polymerisation of propylene and ethylene oxides onto glycerine or trimethylolpropane.

The development of modern polyether polyols can be traced through four main stages. The first commercial development of polyether based flexible polyurethane foams was pioneered (1956–57) in the USA using polyoxypropylene diols of molecular weight about 2000 in a two stage or 'prepolymer' process. The prepolymer process (1957–61) was quickly replaced by single-stage or one-shot systems based upon polyoxypropylene

FIG. 2. Rectangular section polyether slabstock from 'Maxfoam' machine.
(Viking Eng. Co. Ltd, Stockport, UK.)

triols of molecular weight about 3000. The low reactivity of the largely
secondary hydroxyl end groups of polyoxypropylene polyols led to the use
of ethylene oxide tipping or 'capping' to introduce a proportion of primary
hydroxyl end groups and so increase reactivity. This was particularly
important in developing commercially viable cushion moulding processes
and in establishing the use of polyether triols with molecular weights up to
about 5000 in slabstock foam manufacture. It was soon realised, however,
that obtaining optimum reactivity in the foaming process was not merely a
matter of obtaining the best ratio of primary to secondary hydroxyl groups
but required satisfactory compatibility with the other components of the
foam mix over a wide range of formulations. The use of ethylene oxide was
extended by forming blocks of polyoxyethylene units within the
polyoxypropylene chains in order to obtain polyethers with the required
hydrophilic and surface active properties.

Developments since about 1971 have been mainly aimed at improving
the economics of foam slabstock manufacture by improving the
processability and versatility of the polyols. The development of flat-top
block processes, especially the Foamax, Maxfoam and Varimax processes,
has required polyethers with reaction rates to suit the trough foaming
apparatus (Fig. 2).

The polyether polyols used for flexible foam manufacture today are of
three main types: general purpose slabstock polyols, polyols for hot-cure
moulding and polyols for high resilience foam and cold-cure moulding
(Table 2). There are several other types of polyethers finding increasing use
to obtain special properties or effects in both moulding and slabstock.
'Polymer' polyols or vinyl reinforced polyols made by the polymerisation of

TABLE 2
TYPICAL PROPERTIES OF FLEXIBLE FOAM POLYETHER POLYOLS

Property	Hot moulding polyol	Slabstock polyol	High resilience polyol	'Polymer' polyol
Mean molecular weight	3 100	3 500 to 4 000	4 600 to 6 000	5 000 to 6 000
Hydroxyl value (mg KOH/g)	55	42 to 48	36 to 28	46 to 28
Acid value (mg KOH/g)	0·05	0·05	0·05	0·1
Water content (%)	0·05	0·05	0·05	0·05
Viscosity at 25°C (poise)	5·1	5·8	8 to 14	10 to 15
Specific gravity (25/25)	1·010	1·018 to 1·021	1·02	1·03
Flash point (°C)	> 240	> 240	> 240	> 260
Fire point (°C)	> 260	> 260	> 260	> 260
Ethylene oxide content (% of total alkylene oxide)	6 to 12%	8 to 18%	10 to 20%	10 to 15%

acrylonitrile and styrene *in situ* in a high molecular weight cold moulding type polyether are used in both slabstock and high resilience mouldings. They yield increased hardness or load bearing without significant loss of tensile properties and, especially in cold moulding with modified isocyanates and MDI/TDI mixtures, produce more open permeable foams than unmodified polyols. Minor classes of polyethers include those for hydrophilic foams made with high proportions of ethylene oxide (50% to 70% of total alkylene oxide content) and low molecular weight polyols used in high resilience and cold-cure foam manufacture.

Polyethers for Slabstock
Polyethers for slabstock are typically made by the base catalysed oxypropylation of glycerol to about 75% of the final molecular weight and then finishing in two or more stages using both ethylene and propylene oxides, separately or together. This type of polyether has good compatibility with the other components of the foam mix over the normal range of formulations from about 1·8 to 5·0 parts of water per 100 parts of polyol. Finishing the polyether with both ethylene and propylene oxides yields the required reactivity without the critical catalyst concentration needed for simple 'tipped' polyethers.

    Slabstock polyethers are made in molecular weights from 3500 to 4000 with a nominal functionality of 3. The effect of the higher molecular weight is a small increase in load bearing but at the expense of some increase in compression set, reduced resilience and reduced stability during foaming especially of high density and high fluorocarbon blown foams. The lower hydroxyl value of the 4000 molecular weight polyols (OH $= 42$ mg KOH/g) compared with those of 3500 molecular weight is sometimes claimed to give a significant cost reduction but statistical comparisons over weeks of commercial foam production have failed to substantiate real savings because the lower molecular weight polyols tend to give marginally more uniform and lower density slabstock at a given ratio of water/TDI/polyol. Recent practice has seen some usage of polyols of intermediate molecular weights (3600 to 3700) often with a significant diol component giving an effective functionality of about 2·8.

Polyethers for Hot-cure Moulding

The hot-cure moulding process required a polyether polyol with a reactivity designed to give good flow of the expanding foam mix with rapid surface and interior foam cure. This is obtained by partial block polymerisation of ethylene oxide onto a polyoxypropylene triol, followed by a degree of ethylene oxide tipping, combined with a choice of molecular weight between 3000 and 3300. The overall reaction rate of a polyether triol can be adjusted by ethylene oxide tipping but is increased by reducing molecular weight. The use of a lower molecular weight polyether than that used for slabstock gives a fast final cure at the lower temperatures found in moulding without the penalty of reduced flow and sensitivity to closed cells resulting from excessive 'tipping'.

High Resilience or Cold-cure Moulding Polyols

High resilience or cold-cure moulding polyols are triols of molecular weights from 4500 to 6500 with high levels of tipping giving primary hydroxyl levels greater than 60 % of total hydroxyl ends. The effective functionality of these polyols becomes significantly less than three with increasing molecular weight because of the molar effect of side reactions yielding diols and monols.

The optimum foam properties are obtained by polyether triols in the molecular weight range from 5000 to 6000 and primary hydroxyl levels of 75 % to 90 % obtained with ethylene oxide tipping using total ethylene oxide contents from about 10 % to 20 % depending on the molecular weight and precise method of manufacture.

Polymer Polyols

The commercial production of these polyethers has been led by Union Carbide Corporation.[1] They are made by the *in situ* polymerisation of a vinyl monomer in the presence of a conventional polyether polyol. Early products used acrylonitrile alone but these have largely been replaced by copolymer mixtures which give better colour and odour. Standard products such as NIAX 34-28 (Union Carbide) and Desmophen 1920 (Bayer) contain about 20 % of an *in situ* copolymer of acrylonitrile and styrene to give a product with an overall OH value of about 28 mg KOH/g[1]. Increasing vinyl polymer level gives an almost linear increase in foam hardness[2] without significant loss in tensile properties but the viscosity of the polyol increases rapidly with increasing vinyl polymer content so limiting the useful level to about 20 %. Higher levels of modification are possible using methyl methacrylate acrylonitrile copolymer modified polyols and such products have been offered in Europe.

ISOCYANATES

Flexible foam manufacture was first based upon lightly branched polyesters and TDI 65 but since about 1960 the use of TDI 80 has dominated flexible foam manufacture made from either polyether or polyester polyols. The use of TDI 80 is still expanding at a rate near to 8 % each year requiring at least one new 40 000 to 50 000 tonne capacity plant every year. In addition to TDI, modified TDI, modified MDI and mixtures of MDI and TDI have found increasing use in flexible and semi-rigid foam manufacture since about 1970. The principal isocyanate variants in use today are tabulated in Table 3 although many other mixtures of TDI and modified TDI with polymeric MDI compositions have been used for 'cold-cure' moulding, especially in the furniture industry.

The manufacture of TDI by the established route:

TABLE 3

PRINCIPAL TDI AND MDI VARIANTS USED IN FLEXIBLE FOAMS

Isocyanate	Functionality (NCO groups per molecule)	Applications
TDI 80	2·0	Flexible foam (slab amd moulded)
TDI 65	2·0	High load-bearing foams, polyester and polyether slab
TDI–urethane modified	2·15	High resilience, self-skinning foams
TDI/TDI trimer solution	2·3	High resilience foams (slab and moulded)
TDI–allophanate modified	2·2	High resilience foams
TDI mixed with undistilled TDI ('crude TDI')	2·1 to 2·3	High resilience foam (rigid foam), semi-rigid foam
TDI mixed with polymeric MDI	2·1 to 2·5	High resilience 'cold-cure' foam
TDI/polyamine adduct	2·5	High resilience 'cold-cure' foam
Polymeric MDI/modified TDI	2·15 to 2·3	High resilience 'cold-cure' with high load-bearing
Polymeric MDI	2·7 to 2·8	(Rigid foam.) Semi-rigid foam
Urethane modified pure MDI	2·0	Self-skinning moulding, shoe soling
Uretonimine modified pure MDI	2·2	Self-skinning foams
Uretonimine/urethane modified pure MDI	2·1	RIM moulding auto parts
Urethane modified MDI compositions	2·3	Self-skinning, auto parts, flexible foam moulding
Low viscosity liquid polymeric MDI	2·6	High density flexible foam, carpet backing

is described in many publications.[3] A possible alternative route avoids the use of phosgene by the direct reaction of nitro compounds with carbon monoxide[4] and the commencement of pilot plant manufacture by this route was recently announced in Japan.[5]

Modified TDI is used in the manufacture of high resilience foam slabstock and moulding. TDI modified by reaction with a diol to yield a product containing a minor proportion of linear polyurethane has been used mostly in moulding 'cold-cure' type furniture foams. The main product of this type used in Europe was TDI, modified with polyoxyethylene diols and blended with polymeric MDI to give cold-moulded furniture cushions of improved tensile properties at low densities. More popular isocyanates of controlled functionality are made by several methods:

(a) TDI containing TDI trimer (Fig. 3(a)) in solution.

(b) TDI modified with low molecular weight polyols (diols and triols) and heat treated to yield allophanate (Fig. 3(b)).

(c) TDI modified with undistilled TDI ('crude' TDI).

The 'crude' TDIs consist essentially of blends of normal distilled TDI 80 with undistilled TDI or TDI still residues to give a controlled reactivity and functionality. The polyfunctional species are complex molecules arising from the phosgenation of urea by-products.

(d) TDI modified by the addition of polymeric MDI.

Branching introduced by adding polymeric MDI is less effective because of steric factors which inhibit reaction with high molecular weight polyols especially in the presence of water. Polymeric MDI typically has a functionality of 2·7 to 2·8, the composition usually consisting of about 50 % diisocyanatodiphenyl methane (mostly as the 4:4′ isomer, although the 2:4′ is always present at levels from about 2 to 12 % depending on the method of manufacture). The other constituents in order of importance are triisocyanate and up to 25 % of higher functionality polymeric polyisocyanates. Mixtures of TDI 80 and polymeric MDI are used over the range from 20/80 to 60/40 MDI to TDI ratio giving theoretical functionality from about 2·15 to about 2·45. In practice the higher levels of MDI tend to yield softer foams with reduced tensile properties but need only low catalyst levels to give fast curing foams. The use of TDI 65 instead of TDI 80 is common with high MDI to TDI ratios because it gives foams with higher load bearing and reduced closed-cell content.

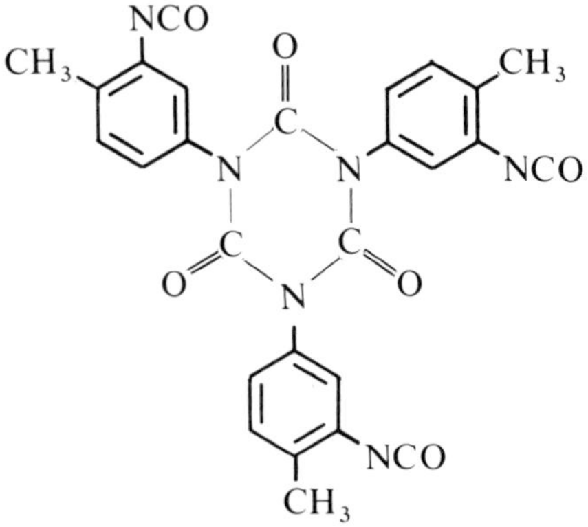

(a)

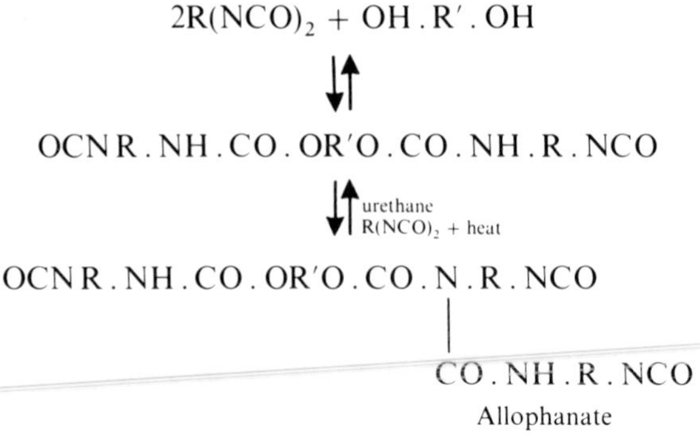

$$2R(NCO)_2 + OH.R'.OH$$

$$\updownarrow$$

$$OCN R.NH.CO.OR'O.CO.NH.R.NCO$$

$$\updownarrow \text{urethane} \atop R(NCO)_2 + \text{heat}$$

$$OCN R.NH.CO.OR'O.CO.N.R.NCO$$
$$|$$
$$CO.NH.R.NCO$$
Allophanate

(b)

FIG. 3.

An adduct isocyanate made by reacting TDI 65 with a polyamine is now being offered in the USA for the manufacture of high resilience foam.[6] The adduct, of functionality about 2·5, is claimed to give increased foam hardness and rapid gelation.

The use of conventional polymeric MDI compositions without TDI now dominates semi-rigid foam moulding for fascia pads, other energy absorbing safety padding and interior trim in European automobiles because, unlike the earlier TDI based systems, one-shot foam systems can be formulated to meet a wide hardness range with good flow and adhesion to the ABS/PVC foils used in auto trim. Polymeric MDI compositions with average functionalities from 2·7 to 2·8 are also widely used in self-skinning foam trim for fascia pads, steering-wheel centre pads and other light duty trim applications. For heavier duty such as steering-wheel rims, arm-rests and bicycle saddles requiring high abrasion resistance and high flexibility, lower functionality MDI variants are used. Uretonimine modified pure MDI is a low viscosity liquid which can be handled easily on high pressure impingement mixing dispensing machines. It has been widely used for self-skinning trim although it has now been superseded for many trim applications by the cheaper and more easily handled 2·3 functionality MDI compositions based on polymeric MDI. Uretonimines remain an important constituent in liquid MDI variants used in the manufacture of car bumpers and external trim.

Uretonimine modified pure MDI is made by heating distilled pure MDI in the presence of a suitable catalyst. The reaction can be represented:

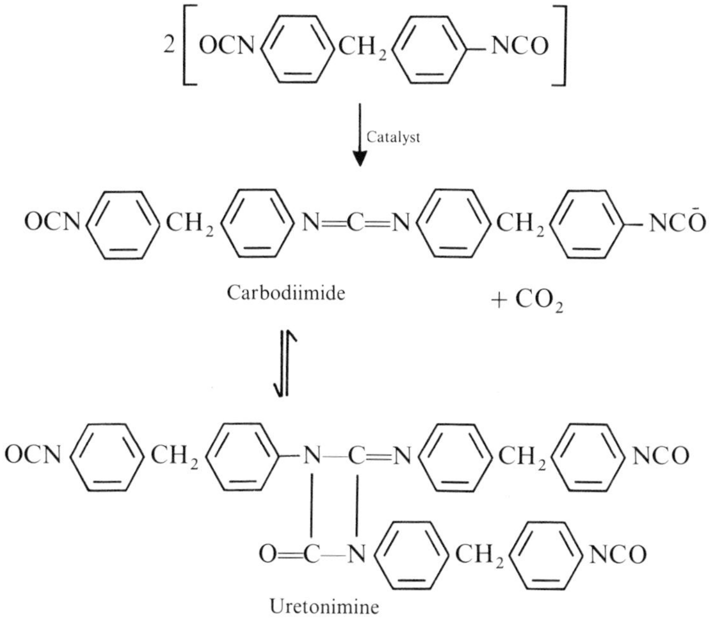

$$2\left[OCN\underset{}{\bigcirc}CH_2\underset{}{\bigcirc}-NCO \right]$$

↓ Catalyst

$$OCN\underset{}{\bigcirc}CH_2\underset{}{\bigcirc}N{=}C{=}N\underset{}{\bigcirc}CH_2\underset{}{\bigcirc}-N\overset{..}{C}O$$

Carbodiimide $+ CO_2$

$$OCN\underset{}{\bigcirc}CH_2\underset{}{\bigcirc}-N-C{=}N\underset{}{\bigcirc}CH_2\underset{}{\bigcirc}NCO$$

$$O{=}C-N\underset{}{\bigcirc}CH_2\underset{}{\bigcirc}NCO$$

Uretonimine

Uretonimine modified liquid MDI is available from all the major suppliers and the various products differ mostly in dimer content and 2:4′ MDI content giving some significant differences in low temperature stability.

Many other MDI based products contain some uretonimine formed as a by-product during heat treatment.

MDI compositions with an average effective functionality of 2·3 are available from a number of suppliers and these products differ significantly in chemical composition depending on the method of manufacture. The effect of these differences on the final product is insignificant but during processing there are noticeable differences in liquid stability and reactivity. Liquefaction of pure MDI is also commonly attained by urethane formation with low molecular weight diols. Very stable liquid products are available with effective functionalities near to 2·0 giving high-grade elastomeric products. The disadvantage of urethane modification as a means of liquefaction is the relatively high viscosity obtained (8 to 16 poise) which make the compounds unsuitable for some RIM applications. The usual compromise is to use a mixture of uretonimine and urethane modification to yield liquid MDI compositions with an effective functionality about 2·1 and a viscosity about 1·5 to 2·0 poise at 25 °C which is suitable for RIM operation, even for large parts.

ADDITIVES: BLOWING AGENTS

Chlorofluoromethanes 11 and 12 (CFMs 11 and 12) are the most used blowing agents in polyurethane foam manufacture, but their use is threatened by agencies, especially in the USA, who wish to ban the use of CFMs. The environmental lobby is concerned about the possible cumulative effects of CFMs on the stratospheric ozone layer which absorbs ultra-violet light with wavelengths between 290 and 320 nm (UV-B). A 1·0 % depletion of the ozone layer is postulated to cause a 2 % increase in the amount of UV-B reaching ground level and the associated effects.[7,8]

The theory is that CFMs 11 and 12 accumulate in the atmosphere at a rate close to their release rate because they are inert and there are no known natural destruction processes. They are transported upwards to altitudes above 25 km where it has been postulated they are decomposed by far UV light to give chlorine atoms which can destroy ozone:

$$Cl + O_3 = ClO + O_2$$

$$ClO + O = Cl + O_2$$

The British Government view[9] is that the moves to ban the use of CFMs is precipitate in view of the tenuous nature of the argument and that the suggested changes in UV radiation must be kept in proportion. They point out that there has been a 5 to 10 % increase in ozone since the 1920s when measurement began and that the amount of UV radiation reaching the Earth's surface varies by a factor of 100 times between latitude 30° and latitude 50°. They conclude that UK policy should be to wait to assess the problem in the light of the extensive research being undertaken and in the meantime it will encourage CFM makers and users to seek alternatives and minimise leakages.

The main uses of CFMs are as aerosol propellants and refrigerants and their use in polyurethane foams amounts to only about 5 % of total CFM production. The use of CFMs in closed-cell rigid foams is an essential feature of the high insulation value of these foams because of the low conductivity of CFM vapour compared with air and carbon dioxide. CFMs used as auxiliary blowing agents in flexible foams and self-skinning trim, however, are released immediately into the atmosphere and there is a risk, therefore, that the use of CFMs in flexible foam systems might eventually be banned or controlled. Although the calculated ozone depletion due to CFMs involves considerable uncertainties, the US Environmental Protection Agency is proposing a ban on CFM usage in non-essential applications. The proposed ban includes CFMs 11, 12 and 22 ($CFCl_3$, CF_2Cl_2 and CHF_2Cl) and is primarily aimed at their use in aerosols, but legislation against some more essential uses has been proposed and has prompted the search for alternatives. In flexible foam, methylene chloride has been used to a minor extent as an auxiliary blowing agent for many years, but its relatively low volatility and its high solubility in the polyurethane polymer limit its usefulness. In slabstock foam manufacture, methylene chloride can be used satisfactorily as an auxiliary blowing agent at levels up to about 4 parts per hundred of polyether and in admixture with CFM 11. A composition of 50 % CFM 11, 30 % methylene chloride and 20 % ethyl chloride is recommended[10] to economise in CFM 11 in making soft and super-soft foam. The increasingly important use of self-skinning foam systems in automotive trim manufacture depends upon the use of CFM 11 as the primary blowing agent, although mixtures with methylene chloride have been used satisfactorily for self-skinning trim for many years.[11] The use of methylene chloride alone is unsatisfactory because of its high solubility giving excessively thick, dense skins with stickiness and poor mould release. It is unlikely that CFM 11 will be banned in self-skinning flexible foams before the early 1980s and little work seems to have yet been done on the development of alternative systems. The best alternative is possibly CFC 123 (CF_3CHCl_2) which has a boiling point of 27 °C, and may have a relatively short life in the lower atmosphere.

Catalysts

A large number of catalysts have been examined and reported but in commercial practice optimum technical requirements are met economically with a small number of products. Slabstock flexible polyether foams are made usually with one of the many grades of stabilised stannous octoate in combination with one or more tertiary amines. The low cost N,N-dimethylethanolamine (DMEA) has been widely used as the sole amine catalyst for low and medium density polyether slabstock but a more powerful polymerisation catalyst is needed with CFM 11 levels greater than a few parts per 100 of polyether and with low water level formulations. Combinations of DMEA with diaminobicyclooctane (DABCO) at concentrations from about 0·01 to 0·15 parts per 100 of polyether are often used in slabstock foam.

Bis-(β-N,N-dimethylaminoethyl) ether (A1) is becoming widely used, often in combination with DABCO, because its balanced acceleration of both the polymerisation and water blowing reactions gives improved control of rising foam flow in both hot and cold-cure moulding.

Cold-cure moulding systems using mixtures of polymeric MDI and TDI were first developed using triethylamine as the main catalyst at relatively high concentrations (ca 0·5 parts per 100 of polyether) to obtain the required high cure rates with a sufficiently open cell structure, but the strong smell and the limited range of load-bearing properties obtained with this catalyst led to its early replacement. N,N,N',N'-tetramethylbutane diamine was proposed as sole catalyst for polymeric MDI based foams[18] but this catalyst is no longer used in Europe and the combination of A1 and DABCO originally suggested[19] in 1970 is commonly used today, sometimes with the addition of DMEA or N-ethyl morpholine (NEM) to increase the flow of the rising foam.

Organo-tin catalysts are less commonly used in polyester foam manufacture which relies largely on amine catalysts such as dimethylcyclohexylamine (DMCHA), dimethylaminocetylamine (DM16D), and methyldicyclohexylamine (MDCHA), with lower activity 'blowing' catalysts such as NEM and dimethylaminobenzylamine (DMBA) to control the rise profile and closed-cell content.

Stannous octoate is the most widely used organo-metal catalyst but its sensitivity to hydrolysis and oxidation prevent its use in blends of polyol containing water and amine catalysts. Dibutyl tin dilaurate (DBTDL) is sufficiently stable for use in such blends although dibutyl tin dialkylmercaptides and dialkylthioglycollates are significantly more stable in slightly acidic water containing polyols.

The use of diisocyanate trimerisation catalysts such as tris(dimethylaminopropyl)-sym-hexahydrotriazine is of value in making slabstock foams with improved resistance to ignition.

Silicone Surfactants

Nearly all 'one-shot' flexible foam processes depend upon the use of silicone surfactants which not only stabilise the rising foam but, by assisting in the emulsification of the water phase in the polyol and isocyanate mix, modify the reaction profile and the properties of the polymer. The silicone surfactant lowers the surface tension of the foam mix so improving the dispersion of nucleating air and helping to maintain the gas in a fine dispersion until the foam is self supporting.

The silicone surfactants are polysiloxane—polyoxyalkylene block copolymers—and the commonest structures used are:

1. Linear siloxane with pendant polyether groups

2. Linear siloxane polyether copolymer

3. CH_3Si

The first stage in the synthesis of silicone surfactants is the direct reaction of silicon and methyl chloride to produce a mixture of methyl chlorosilanes of which the four most important are:

1. $(CH_3)_2SiCl_2$ ca 75 % of monomer yield—chain extender
2. CH_3SiCl_3 Introduces branching
3. CH_3HSiCl_2 Introduces the reactive SiH group into the siloxane chain
4. $(CH_3)_3SiCl$ Chain stopper

The chlorosilanes are separated by distillation and hydrolysed with water yielding silanols which condense forming siloxanes:

$$2{\equiv}Si{-}Cl + 2\,H_2O \rightarrow 2\,HCl + 2{\equiv}Si{-}OH \rightarrow {\equiv}Si{-}O{-}Si{\equiv} + H_2O$$

Cohydrolysis of a mixture of chlorosilanes produces a mixture of siloxanes including low molecular weight cyclics formed from the dichlorosilanes. The mixture is equilibrated by heating at 100 °C with an acidic catalyst when silanol condensation proceeds to completion and the Si—O—Si linkages are broken and reformed until the system is stable.

$$(CH_3)_3SiCl + nCl_2Si(CH_3)_2 + mCH_3SiHCl_2 + ClSi(CH_3)_3$$

hydrolysis and acidic equilibration

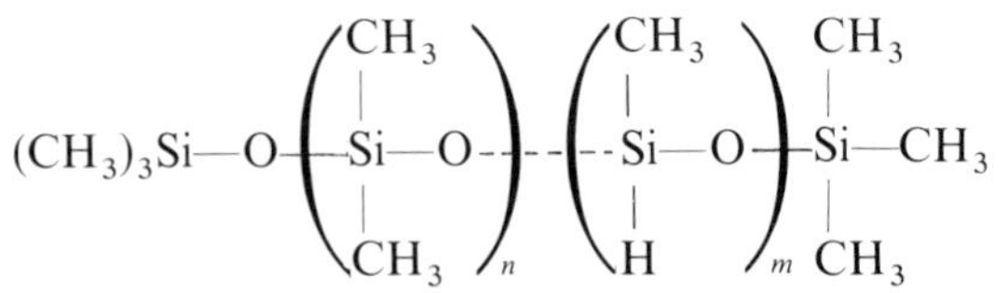

$$(CH_3)_3Si-O\left(\underset{\underset{CH_3}{|}}{\overset{\overset{CH_3}{|}}{Si}}-O\right)_n\cdots\left(\underset{\underset{H}{|}}{\overset{\overset{CH_3}{|}}{Si}}-O\right)_m\underset{\underset{CH_3}{|}}{\overset{\overset{CH_3}{|}}{Si}}-CH_3$$

The commonest alternative route to the siloxane polymer block is:

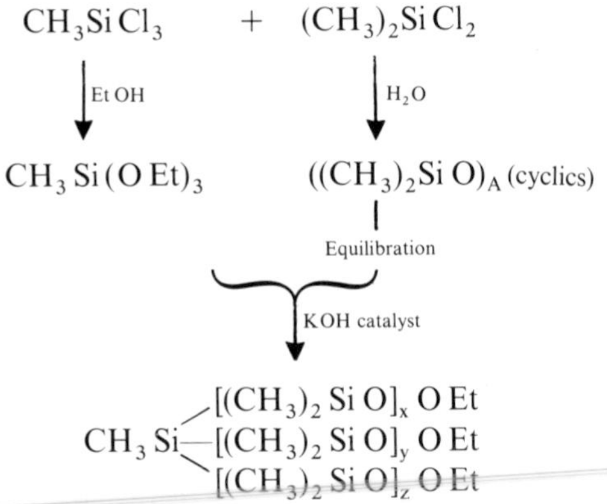

$$CH_3SiCl_3 \qquad + \qquad (CH_3)_2SiCl_2$$

$\downarrow$ Et OH $\qquad\qquad\qquad$ $\downarrow$ H$_2$O

$$CH_3Si(OEt)_3 \qquad\qquad ((CH_3)_2SiO)_A \text{ (cyclics)}$$

Equilibration

KOH catalyst

$$CH_3Si\begin{cases}[(CH_3)_2SiO]_x\,OEt\\ [(CH_3)_2SiO]_y\,OEt\\ [(CH_3)_2SiO]_z\,OEt\end{cases}$$

The ethoxy terminated dimethyl polysiloxane intermediates are reacted with hydroxyl ended polyethers to obtain the desired surfactant.[20]

Non-hydrolysable (Si—C)

$$Si-H + CH_2=CH-CH_2\,OR$$
$$\rightarrow Si(CH_2)_3\,OR$$

Hydrolysable (Si—O—C)

$$Si-OEt + R-OH \rightarrow Si-O-R$$
$$+ EtOH$$

The most widely used silicone surfactants in flexible foam have the general structure:

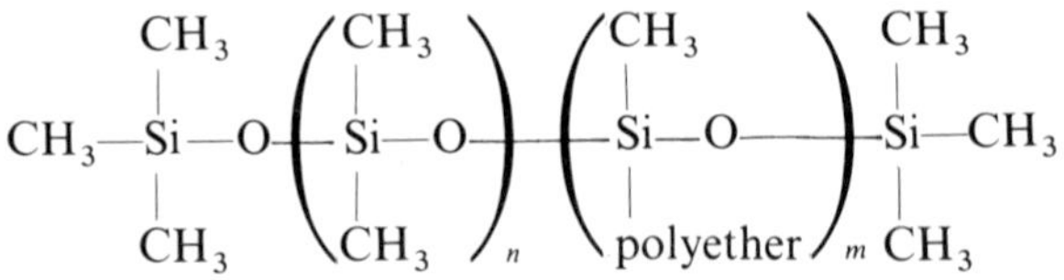

$$CH_3-\underset{\underset{CH_3}{|}}{\overset{\overset{CH_3}{|}}{Si}}-O\left(\underset{\underset{CH_3}{|}}{\overset{\overset{CH_3}{|}}{Si}}-O\right)_n\left(\underset{\underset{polyether}{|}}{\overset{\overset{CH_3}{|}}{Si}}-O\right)_m\underset{\underset{CH_3}{|}}{\overset{\overset{CH_3}{|}}{Si}}-CH_3$$

where the pendant polyether groups are ethylene oxide–propylene oxide adducts containing about 50% ethylene oxide often as a capped polyol.

Fillers

The use of inorganic fillers in flexible foams for seating has not found significant use in Europe except for applications such as sound absorption where mass is important. Inorganic fillers yield foams of increased density and hardness, but reduce tensile strength, tear strength and elongation at break. Better foam properties are obtained by the use of fibrous fillers such as 0·5 to 1 mm chopped polyamide or polyester staple, which at levels as low as 5·0% on total foam weight will yield improved tensile properties combined with both higher hardness and a more linear stress–strain curve.[53] Foam reinforced with $\frac{1}{8}$ in chopped fibre-glass strand was also shown[54] to have increased load bearing, energy absorption on compression and an improved SAG factor.

DEVELOPMENT OF SEATING FOAMS

Testing Methods

The decade (1961–70) following the introduction of polyether based foams saw the rapid development of both foam chemistry and production technology aimed at reducing the cost of upholstery foams whilst improving their physical properties. The conventional physical property measurements were derived directly from those developed for rubber latex foam seating but these specifications, covering density, tensile strength, elongation at break, load bearing and compression set at 50% compression, proved insufficient to characterise polyurethane foams. The versatility of the polyurethane foam process and the pressure on production costs led to the rapid development of very low density foams with good tensile and high load-bearing properties, but many such foams proved unsatisfactory in service showing rapid softening and loss in height. The first attempts to find a test which predicted service performance led to specifications requiring more severe compression set testing at 90% compression and then to static fatigue and dynamic fatigue tests such as the Ford roller shear fatigue test (Ford designation BO-12-2).[12]

National and international fatigue tests have now been established and correlated with service performance so that in many countries the purchaser of seating grade foams can specify the foam according to the severity of the application. The comprehensive range of applications listed in the 1977 revision of BS 3379 typifies five types of service:

Class	Type of service	Application
X	Extremely severe	Heavy duty contract seats
		Heavy-duty public transport seats
V	Very severe	Public transport seats
		Cinema and theatre seats
		Contract furniture seats
S	Severe	Private and commercial vehicle seats
		Domestic furniture seats
A	Average	Private vehicle seat-backs and arm-rests
		Domestic furniture backs and arm-rests
L	Light	Padding, pillows, scatter-cushions

The classification depends on performance in the constant load pounding test described in BSI DD 31:1974 which will shortly be issued as a full British Standard and will correspond to International Standards Organisation method IS 3385:1975 and substantially to Deutsche Normen DIN 53574. The sample is repeatedly loaded and unloaded with a flat circular indentor exerting a force of 750 N at a rate about 70 cycles per minute for 80 000 load cycles and the loss in indentation hardness is then measured. The classification allows up to $12\frac{1}{2}\%$ loss in hardness for class X increasing up to 45 % loss for class L material. The hardness loss by this procedure correlates well with the service performance of the foams in daily use for seating over about nine months.[13]

A classification covering six grades of use, four based upon performance by the roller shear fatigue test and a secondary performance classification based upon fatigue in a static load fatigue test is defined by the American Society for Testing and Materials in ASTM D.3453-76. This standard suggests a labelling procedure which gives the performance grade established by the test procedure.

There is no doubt that the use of dynamic fatigue tests to define service performance has been successful in virtually eliminating the sale of unsatisfactory low density high hardness foams for seating applications and in regaining customer confidence.

Automotive Seating

The automotive industry itself has always tried to correlate testing specifications with service performance. The usual approach is to specify minimum levels for density, tensile properties and compression set together with a dynamic fatigue test procedure to simulate arduous service. Typical basic industry specifications are listed in Table 4. Although the various dynamic fatigue tests used cover a range of loading, frequencies and duration, the results in practice are similar and most of the tests can be met by a foam which meets the class V classification of BS 3379 or IS 3385 and the increasing acceptance of the relevance of this test is indicated in Table 4. All the major automobile makers have basic service tests which they have correlated with the chosen accelerated fatigue tests. There is usually also an accelerated test procedure on complete seat assemblies, often established before the advent of polyurethane foam, to measure seat durability. Procedures such as the 'jounce and squirm' machine of General Motors,[14] the cyclic rocking machine of Leyland, the roller-shear test for covered cushions developed by Ford and the 60 day cyclic test used by Volvo[15] have all been related to the service performance recorded during model durability trials. The General Motors tests, for example, are related by experiment to service performance in a 36 000 miles test around their proving ground using a fully trimmed assembly. There is an increasing confidence in cushions tested by accelerated fatigue tests of the IS 3385 type but the service tests of fully trimmed seats are still necessary to test the assembly design and to reveal frame and seat suspension problems.

Transport seating is required not only to provide a comfortable and durable support, but to augment the vehicle suspension system in absorbing bumps and vibrations generated from the road, engine and transmission. Many studies have shown the harmful effects of vibrations on man[21–23] and the gastro-intestinal disturbance and back disorders investigated are of particular importance to those who travel in vehicles. The car-makers have studied the dynamic performance of seating and early foam cushion specifications aimed at improving the vibration damping of seat cushions include the high hysteresis latex foam used by Saab and the low resonance factor requirements applied to polyurethane foam cushions by Peugeot in 1972.

Polyurethane 'deep seat' cushions, in which the cushion is mounted on a rigid seat frame directly attached to the car body, can form an integral part of the suspension system of the vehicle and there have been many studies on the dynamic properties of such seats both alone and in combination with studies of the mechanical impedance of the human body. Polyurethane flexible foam cushions behave as a damped spring under an applied cyclic displacement and tend to resonate at a frequency around 5 Hz. Cushions are usually evaluated by mounting them on a seat base carried on a

TABLE 4
TYPICAL AUTOMOTIVE H/R SEATING FOAM SPECIFICATIONS

	Make							
	Ford	*Peugeot*	*Citroën*	*Leyland (Austin–Morris)*	*Chrysler (UK)*	*GM (Fisher Body)*	*Audi-NSU Volkswagen*	*Volvo*
Core density (kg/m³)	35 min	35	35	By agreement (but usually 34 minimum)	50 (min)	To drawing or 38 to 43	50 + 10 / − 5 or by agreement	45-50
Compression set (22 h at 70°C)								
50% compression	10% max	10% max		10% max	10% max	20% max	8% max	
75% compression						20% max		
90% compression								6% (50°C) / (38°C 98% RH)—10%
Tensile strength (kN/m²min)	80	75	110	60	83	82	120	
Elongation (% min)	120	110	150	90	90	150	130	
Tear strength (N/m min)			333		134	260		
Indentation hardness N/200 min Indentor								
at 25%	210 ± 20 N	21			Several ranges, from 120–240 N	To drawing (ca 200 N)	320 ± 20 N	
at 40%								
at 50%		30						
at 65%		42						
Compression hardness Compression 25%			1·9 to 2·3	3·0 to 4·0				
kN/m² Compression 50%			2·6 to 3·3	5·0 to 7·0			5·5 ± 1	
Compression 65%			4·0 to 5·0	10 to 14				
Fatigue test type	[a]IS 3385 (80 000 cycles)	(250 000 cycles)	(250 000 cycles)	[a] (80 000 cycles)	[a]IS 3385 (80 000 cycles)	Roller shear (20 000 cycles)	Audi test 10^6 cycles	[a]DIN 53574
Maximum hardness loss	30%	10%	5%	20%	30%	25% max		10%
Maximum thickness loss	5%	5%		5%	5%		13 mm max	—
Specification reference	SKM 98D 9 660-A	FP 03	LCI 262		M660/2 200·5	FBMS 7-7	TL-VW 567	852 DRWG 1 281 103

[a] IS 3 385 or similar test procedure.

vibrating plate, which is actuated by a variable speed hydraulic piston to give a sinusoidal motion of small amplitude (usually less than 3 mm). The cushion carries a weight of about 60 kg, equivalent to the effective weight of an average driver/passenger. The exact weight necessary to simulate the load in service depends upon the seat height above the floor and the driving position assuming that both hands are on the steering wheel and that the feet and lower legs are supported by the vehicle floor. The frequency of the

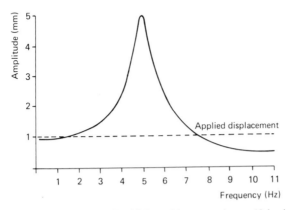

FIG. 4.   Typical resonance curve (3 in thick cushion carrying a 60 kg indentor).

sinusoidal displacement is varied over the range from 0 to 10 Hz or 0 to 20 Hz and compared with the displacement of the load on the cushion. A typical plot showing a resonance frequency about 5 Hz is shown in Fig. 4. This type of test is useful for comparing the dynamic performance of foam cushions and seat constructions but several studies have shown that the resonance frequency of the combination of a human body and a seat cushion is significantly different from the cushion alone.[24,25] Typically, the resonance frequency of a man–seat combination is about 1·5 to 2 Hz lower than that of the seat cushion alone for small amplitude vibrations, although the result may change significantly with the attitude and the degree of relaxation of the driver.[26] The stiffness and damping factor of polyurethane foams tend to increase with increasing amplitude of vibration while the resonance frequency decreases.[25]

    The total effect of the tyres, vehicle and seat suspension on the passenger is often determined by mounting the vehicle on vibrating tables, measuring the effect on the passenger, and relating this ideal relationship to test track results. Measuring the resonance behaviour of seat cushion mouldings alone does, however, provide a quick and simple method of comparing foam systems and significant improvements are possible by adjusting the density and formulation of the foam. Resonance frequency shows only small changes with foam density and stiffness over the range from about 4·5 to 5·5 Hz, the softer foams showing the higher resonance frequency. The resonance amplitude varies from less than four times the applied displacement for stiff low density hot-cure mouldings to about six times for high density high resilience foams, although cold moulding systems can be formulated to give resonance factors similar to those of hot-cure systems. The studies based on sinusoidal vibrations are useful for design purposes and are related to the absorption of engine and transmission vibrations but bumpy roads often give irregular displacements. It has been shown,[27] that the vibrations arising from a bumpy road are absorbed by a deep foam seat to a much greater degree than by a conventional seat, and that no resonance symptoms from this source are detectable under practical conditions.

Trends in Hot- and Cold-cure Moulding
There is an increasing use of moulded flexible foam in European automobiles for both comfort and safety. Moulded units include seat and squab cushions, head-rests, sun visors, sound absorbing underlays and crash padding. The moulding of crash padding has been well established since the early 1960s but the use of thick open-cell flexible foam for sound absorbing underlays which was established with cut slabstock foam has recently been replaced by cold-cure type foam moulded onto a heavy atactic polypropylene backing yielding a greatly increased efficiency of sound absorption, especially at low frequencies. The foam systems used in this type of large thin section moulding are based upon MDI/TDI mixtures and high molecular weight polyethers to give good filling of the mould with short demould times at the low temperature dictated by the use of atactic polypropylene.

Seat cushion moulding falls into three main classes, hot-moulded seats manufactured largely in big semi-automated plants by the large flexible foam-makers, fast demould 'cold-cure' type seating made in-house, and a variety of high resilience foam cushion types supplied to the car-makers as a substitute for latex foams. The broad trend in Europe seems to be towards in-house cushion manufacture by some major car producers who demand fast demould systems for high productivity. Several MDI/TDI systems are in use with mould occupation times of 4 to 5 min. Hot-moulded cushions are made with demould times of 6 to 7 min and lower material costs but the high energy requirements and the effect of the high temperatures used

(170–180 °C) on the working environment are tending to negate the raw material cost advantage. The use of 'vinyl reinforced or polymer polyol' based systems for automotive seating has made little impact in Europe and it now seems unlikely that these products will be used except for cushioning made to meet American specifications.

Mechanical Seating Processes

The concept of making a flexible foam in a preformed cover and capping the rising foam with the seat base to obtain a composite seat in one operation was described in 1959,[28] but since that time many people have developed the concept, mainly aiming at the automotive seating market. The first mass-produced car with moulded deep-seat cushions was the Austin/Morris 1100 of 1962 which was launched with a moulded seat pan and cushion covered with a conventional sewn PVC leathercloth cover. Most cars now use moulded seat cushion or topper pads because of their advantages in performance and dimensional reproducibility but the replacement of the labour intensive cut and sewn cover has made slow progress although several technically satisfactory processes exist. The simple replacement of the sewn cover with a vacuum-formed ABS/PVC skin has been used on several vehicles (e.g. Peugeot 504 and Range Rover) but the complete mechanical seat where foam is made in the preformed cover has been used mainly in commercial and agricultural vehicles. Bedford vans, farm tractors and motor cycles have been the main market for the slush-moulded or in-mould sprayed, PVC plastiscol cover fitted with a relatively high density hot-cured foam. Plastiscol covers, however, are unsuitable for passenger vehicles because they are sticky and uncomfortable in hot weather and soil easily. An alternative process developed in 1973–75 by ICI Paints Division used PVC powder coating of a heated metal mould which was subsequently filled with hot-cured foam. These integral cover cushions had good durability and offered great design freedom both in seat contour and surface pattern. The process could not be adapted to follow the fashion swing to textile covers. Most systems offered today use stretch fabrics usually knitted nylon backed with a low permeability formable foil often combined with a peeled foam or blownlayer. Experimental seats have been made using combinations of fabric with blown polyethylene film, thin thermoplastic polyurethane rubber film, blown latex (SBR) sheet and laminates of blown and unblown PVC.

Probably the most developed process in Europe is the 'Controform' process which has been developed by major automotive seat suppliers in England, France, and Italy using the PVC/fabric laminates and forming processes of the licensor (Storey Bros., Lancaster, England). The laminate is vacuum-formed and placed in a foaming mould where it is held by vacuum and filled with specially developed cold-cure foam. Combinations of foam formulations and cover characteristics have been made to meet the major automotive specifications and to meet all known and proposed safety requirements. Energy absorbing foam seat-backs have been combined with resilient foam squabs to protect the rear passenger.[29] Such Controform seats have satisfactorily completed service testing over 100 000 miles in police cars and commercial vehicles.

The foam systems used are based upon MDI/TDI mixtures to give true cold-curing foams with good adhesion to the formed cover. The properties of the foam system are matched to the cover construction to give maximum comfort.

An alternative single stage process is the 'Skinform' process developed in Italy and used over several years for small series production vehicles. This process avoids the vacuum-forming step, the cover being stretched by vacuum, with plug assist, into the foaming mould and filled with a cold-moulded foam. The foam must have good tensile strength and adhesion to the cover to ensure good service performance. The preferred fabric construction is a multi-layer of fabric/peeled foam/PVC foil or thermoplastic polyurethane foil/peeled foam. The inner layer of peeled foam ensures good adhesion to the foam while the intermediate foam layer gives good ventilation and a soft handle.

The principal features of the processes are covered by patents.[30]

HAZARDS IN FLEXIBLE FOAM MANUFACTURE

Chemical Hazards

None of the chemicals normally used in flexible foam manufacture are highly toxic but diisocyanates are reactive chemicals and need to be handled with care. The toxicology of foam chemicals is described in detail in Chapter 10. The main practical hazard is the respiratory irritant and sensitisation effects of the vapour. This has resulted in legislation fixing a maximum permissible concentration in the atmosphere to which people may be exposed, the Ceiling Threshold Limit Value, TLV(C), at 0·02 ppm (v/v). It is defined as the maximum concentration in the atmosphere that can be tolerated throughout a 7 to 8 h working day or a 40 h working week. It is expressed in ppm, i.e. parts of vapour per million parts of air by volume at $25\,°C/760\,mm$ and in milligrams of vapour per cubic metre of air.[33,34]

There is little difficulty in handling and storing TDI providing the recommended methods[16,31,32] are followed.
The vapour hazard is greater in the foam manufacturing process because some TDI may be vaporised in the exothermic foam-making reaction. A large flexible foam slabstock plant will be almost completely enclosed in a ventilated tunnel. The first stage where the foam rises is where most TDI and catalyst vapour is released, and it is connected to an extractor fan rated between 10 000 and 20 000 cubic feet per minute depending on the rate of foam-making. Satisfactory operation must be monitored by regular measurement of TDI vapour levels but for a known plant condition quick checks of satisfactory function are most easily obtained by measuring air-flow rates into the ventilated tunnel from the operating positions. As a rough guide, air-flow rates of about 150 ft/min through the restricted openings into the tunnel will give safe operating conditions. Large slabstock lines will have at least two further extraction points, one from the curing section of the tunnel and one over the foam block cut-off machine. It is, of course, important to make sure that the extraction system is in balance so that TDI vapour is not pulled down the conveyor. The contaminated air extracted from the foaming area on a large slabstock machine will typically contain about 5 ppm of TDI vapour but the levels in the extract from the curing and cutting areas will be very much less. The blended effluent passing through the extractor fan from several extract ducts usually has a TDI level of one ppm or less and the diluted effluent air is exhausted at a sufficient height for satisfactory dispersal. TDI vapour levels as measured in the extract ducts are fugitive because the air contains a relatively high water vapour level usually with traces of amine catalyst; nevertheless, TDI levels above the TLV may persist for several minutes. An alternative to high level dispersion is to remove the TDI vapour by scrubbing with aqueous alkali. Scrubbers are available to handle effluent air containing several ppm of TDI vapour.[35]

Fire Hazard in Foam Manufacture
The manufacture of large blocks of flexible foam involves a risk of spontaneous combustion if manufacturing conditions are inadequately controlled. The reactions involved in foam manufacture are exothermic:
Primary polymerisation reaction

$$\text{\sim\sim\sim OH} + \text{\sim\sim\sim N}{=}\text{C}{=}\text{O} \longrightarrow \underset{\text{urethane}}{\text{\sim\sim\sim N}\overset{\displaystyle H}{-}\overset{\displaystyle O}{\underset{\|}{C}}{-}\text{O}\sim\sim\sim}$$

Primary blowing reaction

$$H_2O + \text{ww}N{=}C{=}O \longrightarrow \left[\begin{array}{c} H \quad O \\ | \quad \| \\ \text{wN} - C - O - H \\ \downarrow \end{array}\right] + CO_2\uparrow$$

$$\text{wwNH}_2$$

$$NH_2 + \text{ww}N{=}C{=}O \longrightarrow \begin{array}{c} H \quad O \quad H \\ | \quad \| \quad | \\ \text{wwN} - C - N\text{ww} \end{array}$$
$$\text{urea}$$

Clearly the more water and isocyanate are used for CO_2 blowing to make low density foams the more heat is evolved and the higher will be the final foam temperature. It is usual to use 3 to 12 % more than the stoichiometric amount of isocyanate, i.e. an 'isocyanate index' of 103 to 112. The excess isocyanate increases foam hardness by secondary cross-linking reaction with the active hydrogen of the primary urea and urethane linkages to give biurets and allophanates, respectively. These reactions, however, unlike the reaction with water, are reversible and the biurets and allophanates dissociate more easily than the ureas and urethanes. Excess isocyanate groups may also react with water to give amine-ended chains which may further react, at the curing temperatures in the hot block of foam, with primary isocyanate groups or those formed by the dissociation of secondary linkages.

Large blocks of foam 'water-blown' to densities around $20\,kg/m^3$ will reach internal temperatures of about 168 °C and, depending on the size of the block, this temperature may be maintained for over six or seven hours. The danger of spontaneous combustion arises in such hot foams when machinery or formulation faults cause excessive secondary reactions or lower resistance to 'scorch' and fire. Scorch is the yellow/brown discoloration of the centre of large blocks of foam which results from oxidative degradation on exposure to air at temperatures above about 170 °C and should be regarded as a danger signal. Scorch always precedes spontaneous combustion. Since 1970, significant advances have been made in our understanding of scorch and its inhibition by antioxidants. These improvements, however, have only kept step with the increasing size of flexible foam slabs. Many producers accept slight scorch in their high load-bearing low density foams as an indication that they operate the minimum density and therefore most economic formulation.

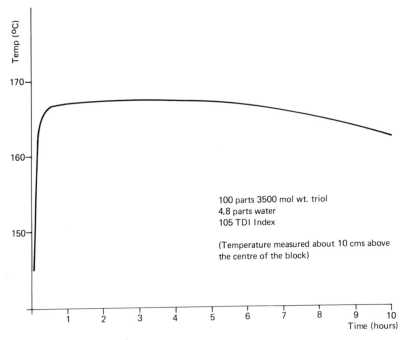

FIG. 5.   Temperature curve of 21 kg/m³ slabstock foam during manufacture.

Safety margins have not increased significantly but risks have been reduced by tighter control of the production machinery coupled with monitoring the machine performance and the foam block curing temperature. The temperature at the centre of a freshly made block of low density foam shows rapid rise during the first few minutes after manufacture followed by a slow increase to a maximum as curing proceeds. The temperature will remain near to the maximum for about six hours depending on the formulation, ambient conditions of temperature and humidity, the size of the block, the cell size of the foam and the method of stacking. A typical temperature curve for a 2 × 1 m section block of 21 kg/m³ density foam is shown in Fig. 5. Polyether polyols supplied for flexible foam manufacture contain antioxidants, of which the principal one is usually 2:6-di-tertiary butyl-4-methyl phenol. At temperatures reached in low density slabstock manufacture, however, antioxidants inhibit the onset of oxidation only in correctly formulated and processed foam. Deviations from the norm which result in higher temperatures may cause 'scorch'.

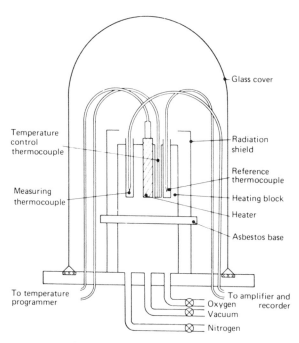

FIG. 6. Arrangement of a differential thermal calorimeter to measure the exothermic oxidation of foam.

The effect of antioxidants in delaying the onset of oxidation can be demonstrated in the laboratory using differential thermal analysis techniques. The standard methods used to evaluate antioxidants in thermoplastics[37,38] can be used to compare antioxidants in polyether polyols. An extension of this approach has been used at ICI[36] to compare the oxidation resistance of actual foams made from polyols containing various antioxidants. A differential thermal calorimeter was constructed in which three cavities in an aluminium block were positioned centrally around an electrical heater (Fig. 6). Two foam samples and an inert material were maintained at 174°C (the temperature at which 'scorch' had been shown to occur in slabstock experiments) in an atmosphere of oxygen. A typical recorder trace of the temperature is illustrated in Fig. 7 which also illustrates the sequence of operations. Thermal equilibrium at the chosen temperature of 174°C is established in an atmosphere of nitrogen, the nitrogen is then evacuated and replaced with oxygen at a flow rate of 10 ml/min. The induction period before the onset of oxidation which is

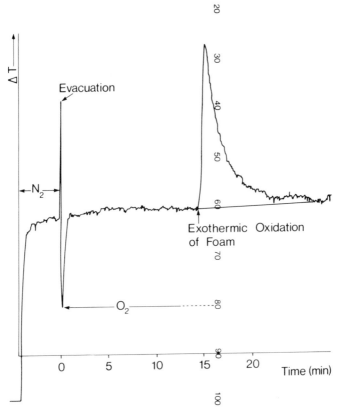

FIG. 7.   Typical recorder trace (174 °C, 10 ml min^{-1} gas flows).

indicated by a rapid temperature rise is a measure of the effectiveness of the antioxidant system. Table 5 shows the induction periods obtained for a number of antioxidants and illustrates the synergistic effect of the widely used combination of 2:6-di-tertiary butyl-4 methyl phenol and pheno-thiazine. Phenothiazine is usually used at a very low level of about 25 ppm of polyether but even at this level it tends to cause some discoloration of the foam. Mercaptobenzimidazole has been shown to be an effective non-staining alternative to phenothiazine and may be used at higher concentrations.

Antioxidant combinations found to be most effective by differential calorimetry were compared in slabstock manufacture on an ICI laboratory slabstock machine. The experiments were done using a high water

TABLE 5
POTENTIAL ANTIOXIDANTS

Antioxidant system		Induction period (min)
No antioxidant added		1
Permanax AN 0·15%		1
Permanax EX		7
Permanax D		1
Permanax SP		10
Permanax NSN		4
Topanol CA		10
Topanol HD		10
Permanax BC		13
Permanax WSP		15
Permanax WSL		8
Permanax HO		10
Vulcafor ZDC		20
Ni dibutyl dithiocarbamate		35
Zn dinonyl dithiophosphate		7
Zn diisopropyl dithiophosphate		22
Phosclere T 315		1
Hilterstab 101		7
Topanol OC	0·05%	10
(2:6-di	0·10%	15
-tertiary butyl	0·15%	15
-4 methyl	0·20%	15
phenol)	0·25%	15
Synergist systems		
0·15% Top. OC 0·375% DLTP		32
0·15% Top. CA 0·375% DLTP		25
0·15% Top. OC 25 ppm Phenothiazine (PTZ)		45
0·15% Top. OC 35 ppm PTZ		40
0·15% Top. OC 50 ppm PTZ		40
0·15% Top. OC 25 ppm Mercaptobenzimidazole (MBI)		35
0·15% Top. OC 35 ppm MBI		40

formulation which was known to yield internal temperatures capable of initiating oxidation scorch in blocks of 1 m square section. Antioxidants were compared at several concentrations in admixture with phenothiazine and mercaptobenzimidazole. The degree of scorch was assessed both visually and by reflective spectrophotometric analysis. The degree of scorch varied from zero to dark brown depending on the antioxidant system used. The best single antioxidant of the hindered phenolic type was confirmed to be 2:6-di-tertiary butyl-4 methyl phenol which gave a reduction in scorch proportional to the antioxidant concentration in the range from 0·01 % (the base level in the special batches of polyether used) to 0·15 % with a small further effect up to 0·2 %. The addition of 25 ppm of phenothiazine gave a marked reduction in scorch at antioxidant concentrations from 0·1 to 0·2 %. Mercaptobenzimidazole was slightly less effective than an equal weight of phenothiazine but higher concentrations could be used without staining.

In another similar series of slabstock experiments the ratios of reactants were varied to simulate machine or metering pump malfunction. The biggest effect on scorch was obtained in foam made with a deficiency of polyol. The maximum recorded temperature inside blocks of foam made with polyether metered accurately (i.e. within one percent of the calculated index, 107) was 168 ± 2 °C in ten experiments. A reduction in polyether flow-rate of 10 % gave an increase in the maximum temperature to 180 ± 2 °C with severe scorching. A further reduction of 10 % in polyether flow 80 % of the normal rate gave an increase in foam temperature to 190 °C, at which point the foam was quenched with water using a water lance.

Accepted methods of preventing scorch and spontaneous combustion involve monitoring the materials, the machine performance and the curing temperature of the foam. The antioxidant level is easily checked by UV spectrophotometry and many large foam-makers measure antioxidant levels in this way although this has already been done by the supplier. The polyol, isocyanate, water and catalyst metering rates should be monitored and recorded individually. Any significant variation from the desired rates requires that the corresponding foam blocks are isolated. In the case of a reduction in polyol flow rate the relevant foam blocks should be immediately removed from the block storage area. As the reduced polyol flow will often cause a noticeable change in foam reaction rate and the block height, this is usually easily marked. The most dangerous condition is an intermittent reduction in polyol flow rate caused by air bubbles which can cause a small volume hot spot within a foam block. Pipe runs and handling procedures must be designed to eliminate the danger of empty feed-lines to the metering pumps which may give rise to air pockets.

Factors which are known to increase the tendency to scorching in border-line formulations are:

(i) Increasing cell count. Unusually fine cells due to excessive nucleating air tend to give higher foam density, a greater surface area and increased scorch.

(ii)   Collapsed foam layers always tend to scorch probably due to a combination of (i) with a reduced rate of cooling.

(iii) Fire retardant halogenated phosphate esters increase the tendency to scorch especially at low concentrations. The effect has been shown to be inversely proportional to the hydrolytic stability of the flame retardant and appears to be caused by acidic decomposition products at the foam curing temperature. Work done on the ICI slabstock machine showed increasing scorch in the series tris-dichloropropyl phosphate, tris-β-chloroethyl phosphate, tris-dibromopropyl phosphate. At concentrations above about 10 parts per 100 parts of polyether the tendency to scorch is reduced because the flame retardant acts as a heat sink and lower maximum temperatures are attained.

(iv)   Increased scorch has been reported[39] from the use of acidic trichlorofluoromethane which has been stored in the presence of water.

(v)   High component temperatures will yield higher foam temperatures causing high scorch and fire risk in sensitive formulations. An increase in polyol temperature from 22 °C to 30 °C, for instance, will cause an increase from slight to unacceptable scorch.

Other factors which have been reported to cause increased scorch in polyether based foams are:

(a)   The presence of metallic impurities, especially traces of nickel, cadmium, chromium, and iron. Work at ICI failed to detect any effect on scorch from adding soluble metallic salts to the foam mix. Work with synthetic mixtures is not, of course, conclusive and many producers use demineralised or distilled water for foam-making to exclude possible contamination from this source. The heavy metal content of the polyol and isocyanate is very low.

(b)   The presence of hydroperoxides in polyols at abnormally high levels has been reported[40] to cause severe scorch as well as severe processing problems which were not reduced by adding further antioxidant of the hindered phenolic type. At the levels found in polyols of normal manufacture, i.e. not more than a few parts per million, variations in hydroperoxide level do not have any detectable effect.

(c)   The practice of adjusting the TDI index to maintain a constant foam hardness with varying ambient humidity, which is common in the areas of northern Europe and North America having severe winters, can increase the risk of scorch. Under low humidity winter conditions excess isocyanate groups tend to react to form allophanate linkages, at curing temperatures between 120–140 °C, and so increase the stiffness of the polymer. This stiff high load-bearing polymer is difficult to make in summer when the air contains much higher levels of water vapour. Under these conditions excess isocyanate groups tend to react with water to form amine end groups which make a lower contribution to polymer stiffness via hydrogen bonding. The usual method of correcting the hardness is by increasing the TDI index so providing more isocyanate groups to react with water and the resultant amine groups. The net effect is to increase the exothermic heat evolved during the curing stage and so increase the tendency to scorch and the risk of combustion. Clearly it is advisable to set foam specifications and formulations which can safely be met under humid conditions.

(d)   Low stannous octoate catalyst levels and the use of low activity stannous octoate compositions have been reported as a cause of increased scorch. It seems likely, from work done in ICI, that the use of insufficient catalyst only indirectly affects scorch by yielding foam of higher density and air permeability.

Foam Handling
Once the foam has been made it is transferred through a fire-wall to a ventilated block curing area where the foam is kept for 12 to 18 h until curing is complete and the foam has cooled. Blocks are usually stored horizontally and must be separated by an air-space of at least 20 cm. Blocks of foam longer than 2 m cured in a vertical position tend to scorch in the centre of the upper half of the block to a significantly greater degree than blocks stored horizontally. Many producers isolate the first and last block of foam from each production run because of the higher risk of excess air and off-ratio metering especially at the start of the run.

Selected blocks of foam are monitored by inserting low heat capacity thermocouples to measure the temperature just above the centre of the block. The temperature usually begins to fall after about 4 h. Blocks which retain their maximum temperature much longer than this are likely to show scorch while an increase in temperature above about 174 to 175 °C is a danger sign. Such blocks are best quenched by the insertion of a metal tube through which water can be pumped into the centre of the block.

It cannot be assumed that foam blocks more than 12 h old are completely safe. Two cases are known of spontaneous combustion occurring later than this, at 14 and 20 h after manufacture. This suggests that once a sufficiently high temperature has been reached, a slow exothermic oxidation in the centre of a very large block of the insulating foam can be sustained until equilibrium is disturbed.

Block storage areas should be monitored by both smoke and heat sensors coupled to alarm systems and automatic quenching using carbon dioxide and water.

Once the foam has cooled, all danger of spontaneous combustion has passed. Foam in large quantities, however, presents a serious fire hazard because its very large surface area to weight ratio permits very fast combustion. There should be no smoking in cutting, handling and storage rooms and good housekeeping to prevent the accumulation of scrap foam is essential.

Hazards in the Use of Flexible Foam

Flexible urethane foam has made a tremendous contribution to improving the comfort of upholstered furniture, simplifying its construction and reducing its cost. The foam consisting of 2 % to 3 % by volume of an organic polymer in a network containing 97 % to 98 % of air is essentially flammable. Especially in domestic furnishing, it is often combined with other combustible materials, some of which are fairly easily ignited by low temperature heat sources such as cigarettes and matches.

Since about 1973, increasing attention has been focused on fires involving upholstered furniture. The media have frequently allocated responsibility for fires and fatalities to urethane foams when they were not responsible for the particular fire situation. Rigid urethane foams have been confused with flexible urethane foams, and fire hazards during the manufacture of flexible foam or the storage of large quantities of flexible foam confused with the fire hazards existing when the foam is used in upholstered furniture. Concern about the fire hazards of plastics and other materials is, of course, natural and should be welcomed by the industry representing as it does a challenge to invention and design. The inaccuracies and sensationalism of some of the media and some politicians is, however, to be deplored.

During recent years a great deal of work has been carried out, both in Europe and the USA, to obtain accurate data about the performance in fire of furniture upholstered with urethane flexible foam. Work has been carried out by government laboratories in both the UK and the USA, and a substantial further amount carried out or funded by the industry. Major investigations have been done for the International Isocyanates

Institute (III) for the British Rubber Manufacturers Association (BRMA) and for the Society of the Plastics Industry (SPI) in the USA.[41]

For some years the industry has recognised that a fire in a manufacturing plant or warehouse where large quantities of foam are stored, is a serious and hazardous affair. Tests carried out at the UK Fire Research Station[42,43] have shown that temperatures of 1000 to 1200 °C and a high concentration of carbon monoxide as well as isocyanate vapour and hydrogen cyanide may be produced in a few minutes when stacks of 200 cubic feet of flexible foam are burning. The temperatures found by the UK Fire Research Station in its investigation of bulk storage have been quoted by critics of the use of urethane in upholstery as if they occurred in domestic fires. Clearly this is not so, and data published by the III, the BRMA and the UK Government investigators themselves do not support such statements, nor are isocyanate vapour and hydrogen cyanide a significant hazard in upholstered furniture fires.

The combustion characteristics of any material and especially a fibrous or cellular organic substance, involve many factors such as ease of ignition, smouldering propensity, rate of heat release, etc. Some of the combustion characteristics of flexible urethane foams can be altered by varying the formulations and including flame retardants and other additives in the production. In almost every piece of upholstered furniture, however, the cushioning material is surrounded by a covering material so that it is the latter which is first affected by a source of ignition. Thus, it is essential to examine how the covering materials used in practice can be ignited by typical ignition sources when tested in combination with the most important types of flexible foam, and to study the extent of subsequent fire damage.

Wilson[44] has reported on the large-scale fire tests carried out by the III in 1973. Following preliminary work in the room and corridor facility of the Rubber and Plastics Research Association (RAPRA) in the UK, the III carried out some fully furnished room burns in a fire test building at the Fire Services Technical College, Moreton-in-Marsh, England. A downstairs room was fitted with carpets, wooden shelving, curtains, etc, and a modern three-piece suite. In one set of tests the padding of the suite and the filling of its cushions were conventional polyether polyurethane foam, in the other set the foam was replaced throughout by wool-based flock traditional material known to burn more slowly. Temperatures and gas concentrations were measured both in the room where the fire was started and in the area of the stairs and bedrooms above. These measurements showed that the concentrations of toxic gases and the temperatures observed during the fire were much lower than those ascribed to urethane flexible foam furniture by the media.

Wilson also reported work showing that thermoplastic coverings were more resistant to ignition from small sources such as cigarettes, but melted under the influence of radiant heat generated from a larger fire source. In his paper, the beneficial effect of an interlayer of flame-retarded cotton fabric interposed between the foam and the covering was reported. This interlayer had the effect of reducing considerably the rate of burning of the furniture which in some cases took three times as long to burn compared with those in which the interlayer was absent. Subsequent further work by the III has confirmed the beneficial effect of interlayers and the substantially increased resistance to ignition and to fire spread by their use. The work by the BRMA has compared the effect on many commonly used covering and cushioning materials, the effect of dust, ageing, etc. on these, and the effect of furniture design upon performance in fire. Work by the BRMA has shown that contamination of fabric covers by dust, cigarette ash, etc. can have a wicking effect which reduces resistance to small ignition sources. They confirm the observations of the III that the cover is significantly more important than the interior in determining the ease or difficulty with which upholstered furniture may be accidentally set on fire. Their observations on furniture design are particularly important, showing that vertical surfaces generally burn three times faster than horizontal, and that burn rates in the vertical plane are reduced where the surface is broken by sufficiently wide furrows. They also emphasise the importance of avoiding gaps at the junction of the back and the seat cushions where otherwise a cigarette trap can be created. Work by the BRMA has also categorically refuted the misunderstanding which had been fostered by some spokesmen that flexible urethane foam and the furniture upholstered with it is capable of spontaneous combustion.[45]

The UK Government maintained that they were not prepared to legislate to control the fire properties of upholstered furniture until they had sufficient information about the fire hazards of plastics in general, and urethanes in particular. They placed contracts with the UK Fire Research Station and certain other laboratories which over a period of 3 to 4 years are intended to provide the information base for legislation. A number of reports of the work funded by the UK Government are already available.[46–9] The research work sponsored by the UK Government involves a study of:

1.   The ignitability and burning behaviour of various foams and

fabrics taken separately and in combination when subjected to standard laboratory tests and to *ad hoc* tests using ignition sources such as cigarettes, matches, etc.

2. The ignitability and burning behaviour of individual items of furniture and furnishings when subjected to common ignition sources.

3. The rate of growth and spread of fire, the rate of temperature increase, and the generation of smoke and toxic gases when fully furnished rooms are set on fire either with restricted or full ventilation.

The work is important for two reasons: it will guide the government in devising regulations or Codes of Practice and it marks a fundamental change in the philosophy of governmental fire testing since it depends upon burning full-scale simulations of furnished rooms. Results up to the present can be summarised by saying that furniture of modern construction, i.e. light-weight frames and foamed plastic cushions covered with synthetic fabrics, is more easily ignited and burns more fiercely than heavy furnishings of traditional construction. The Government reports, however, that improvement in the fire properties of modern furniture are technically readily obtainable. In tests the flammability of a modern suite was reduced to that of traditional furniture by the use of foam with less flammable cover fabrics, together with the use of a flame-retardant interlining. (These conclusions are similar to those reached by the III.) Another observation of interest was that traditional furniture smoulders whereas modern products generally do not.

The UK Government has decided that it is not practicable to impose regulatory controls on the sale of furniture until suitable fire test methods are developed. They are now sponsoring a programme to generate and assess new tests.

Meanwhile, the UK Property Services Agency, the government department responsible for major purchases on behalf of Government, are circulating a purchasing specification which includes a fire test procedure which will eliminate bad combinations of foam and fabric[51]. The observation in the UK Government Reports that traditional furniture smoulders whereas modern products generally do not has been supported by the work of Damant in the Bureau of Home Furnishings at the Department of Consumer Affairs of the State of California[50]. In his recent paper, Damant has pointed out that of the estimated 4·6 million residential fires annually in the USA the fabric item is first to ignite in an estimated

469 000 fires including 93 000 fires involving upholstered furniture. Of the estimated 93 000 residential upholstered furniture fires, the ignition sources in 88 % (82 000) of the incidents are cigarettes/cigars/pipes. Damant points out that fires starting as a smouldering combustion in furniture liberate considerable quantities of volatile combustion products. Flaming combustion may or may not be a sequel to smouldering. Damant's work has studied the smouldering characteristics of a representative sample of upholstery fabric when combined with typical furniture, primary substrates or filling materials. The effect of such factors as the nature of the substrate, the fibre content, weight, weave, and back coating of the fabric on the smouldering characteristics of furniture systems are discussed in his report.

Damant agrees that the exterior fabric has an important influence on whether or not a cigarette causes ignition in a furniture system, and reports that fabrics based on cellulosic fibres are the most hazardous in terms of cigarette-induced smouldering combustion. Most thermoplastic fabrics usually performed well in the cigarette tests carried out in this laboratory since they generally did not sustain smouldering combustion, nor did they transfer sufficient thermal energy to induce smouldering in interior filling materials. Damant also observes that flame retardant finishes on cellulosic fabrics are usually not effective in preventing or inhibiting the occurrence of smouldering combustion in the fabrics. The Bureau of Home Furnishings work has shown that cotton batting used as a filling material in upholstered furniture and bedding for many years is an insidious material in terms of smouldering combustion. A self-sustaining smouldering of cotton batting is readily induced by low heat flux ignition sources such as a spark or a lighted cigarette and conventional urethane substrates are significantly more resistant to smouldering than either conventional or fire-retardant cotton batting. Results on fire-retardant urethane flexible foams showed that resistance to smouldering was dependent upon the formulation chosen, some systems being markedly better than others in this respect. Fire-retardant foams containing antimony trioxide and polyvinyl chloride smouldered more easily than non-fire-retarded foams while the best fire-retardant grade was only slightly more resistant to smouldering than non-fire-retardant foams even though all the fire-retardant foams examined met open flame resistance test requirements. Damant points out that it should be realised that addressing one aspect of combustion may in fact aggravate another equally critical aspect.

In the USA the SPI Urethane Safety Group has sponsored and carried out a considerable amount of work on the fire resistance of flexible urethane foam. Their conclusions are similar to those reached in the UK in the work by the III, the BRMA, and the Fire Research Station, and are summarised in the Urethane Safety Group Bulletins U105 and U106.

It is not possible in this chapter to review the many fire test methods developed in recent years for flexible urethane foams and the work done in various countries of the world. Mention should, however, be made of the new regulations in France aimed at controlling the amounts of synthetic materials containing nitrogen and chlorine used in public buildings. A full description of the reasons for the development of the arreté and its technical requirements has been published recently with a summary of the situation in France[52].

REFERENCES

1. PATTEN, W. and PRIEST, D. C. (1971). 15th SPI International Conf., North Carolina, Sept.
2. KURYLA, W. C., CRITCHFIELD, F. E., PLATT, L. W. and STAMBERGER, P. (1966). J. Cell. Plast., 2(2), 84.
3. TWITCHETT, H. J. (1974). Chem. Soc. Review, 3, 209–30.
4. HARDY, W. B. and BENNETT, R. P. (1967). Tet. Lett., 961.
5. Chem. Engineering, 1977, 17 Jan., 67.
6. WOLFE, H. W., JR, BRIZZOLARA, D. F. and BYAM, J. D. 18th SPI Conf., Oct. 1975; J. Cell. Plast., 13(1), 48, 1977.
7. JOHNSTON, H. S. (1975). Ann. Rev. Phys. Chem., 26, 315.
8. 'Halocarbons: their effect on stratospheric ozone', National Academy of Sciences, Washington DC, 1976.
9. BRITISH DEPARTMENT OF THE ENVIRONMENT, Press Notice 456, 7th Sept., 1977.
10. BURT, J. G. and BRIZZOLARA, D. F. 18th SPI Conf., Oct. 1975; J. Cell. Plast., 13(1), 57, 1977.
11. Mobay Technical Bulletin, 'Integral skin formulations based on MONDUR MR', 1968.
12. BALL, G. W. and DOHERTY, D. J. (1967). J. Cell. Plast., 3(5), 223; 3(7), 317.
13. Furniture Industry Research Association Report No. 23.
14. GM Test Method PETM 171 and PETM 104.
15. Volvo test 5. 3. 5, Drawing No. 1251103.
16. Isocyanates in Industry, British Rubber Manufacturers' Association Ltd, 1977.
17. MAGID, E. B., COERMANN, R. R. and ZIEGENRUECHER, G. H. (1960). J. Aerosp. Med., 31, 913–24.
18. Jefferson Technical Bulletin, 'Cold molded flexible foams based on polymeric MDI', 1968.
19. Union Carbide Technical Bulletin, 'Cold molded urethane flexible foams', 1970.
20. PLUMB, J. B. and ATHERTON, J. H. (1973). Copolymers containing polysiloxane blocks, In Allport, D. C. and Janes, W. H. (Eds), Block Copolymers, Applied Science Publishers Ltd, 306–30.

116   G. WOODS

21. POSTLETHWAITE, F. (1944). 'Human susceptibility to vibration', *Engineering*, **157**, 61.
22. POULSON, E. C. (1949). Tractor drivers' complaints, *Minnesota Medicine*, No. 32, 386.
23. POLI, J. P., LAURENT, J., PARTY, A., PORIEL, P., RICHARD, J., SUSBIELLE, P., GRAND, F., DARQUET, P. and SEVIN, A. (1962). Enquète medico-psychologique sur les conducteurs d'engins. Report OPPBTP, Bd Magenta, Paris.
24. DIEKMANN, D., *Automob. Techn. Zeitschrift*, **58**, No. 8, 209.
25. WISNER, A., DONNADIEU, A. and BERTHOZ, A. (1964). *Int. J. Prod. Res.*, **3**, No. 4, 285–315.
26. COERMANN, R. R. (1963). The mechanical impedance of the human body in sitting and standing positions at low frequency, *In* Lippert, S. (Ed.), *Vibration Research*, New York, Pergamon Press.
27. ICK, J. (1975). The application of flexible polyurethane foam for automotive seating, 18th Annual Tech. Conference, SPI, Oct.
28. ALEXANDER, H. and HENRY, A. M., British Patents 844,650 and 873,518 (1959).
29. BABBS, F. 'Design and development of vehicle seating', Conference papers, Bodytech '75.
30. Spanish Patents 377869, 377870, 378027 1970 (Skinform); US 3,204,016 (General Tire); British 1,129,474 (Dunlop); US 3,390,214 (ICI); British 1,026,016 (Bayer).
31. *Urethane Foams; 'Good Practices for Employees' Health & Safety'*, NIOSH, 4676 Columbia Parkway, Cincinatti, Ohio, USA.
32. Isocyanates: hazards and safe handling procedures, ICI Technical Information U93.
33. UK HEALTH & SAFETY EXECUTIVE, Guidance Note EH15/76, *Threshold Limit Values for* 1976.
34. *Threshold Limit Values for Chemical Substances and Physical Agents in the Working Environment with intended changes for* 1977, American Conference of Governmental Industrial Hygienists, PO Box 1937, Cincinnati, Ohio 45201, 1977.
35. 'Cleme' Gas Scrubber. Leaflet No. 655/NS, Viking Engineering Co. Ltd, 1976.
36. HUGHES, J. F. Research Department, ICI Organics Division, Private communication.
37. GARN, P. D. (1965). *Thermoanalytical Methods of Investigation*, Academic Press, New York.
38. SLADE, P. E. and JENKINS, L. T. (Eds) (1966). *Techniques and Methods of Polymer Evaluation*, vol. 1, *Thermal Analysis*, Arnold, London.
39. BOLGER, M. A. Kay Metzeler Ltd, Private communication.
40. HUNTER, J. Vitafoam Ltd, Private communication.
41. HURD, R. (1976). *4th SPI International Cellular Plastics Conference*, Montreal, Nov., p. 183.
42. WOOLLEY, W. D. (1972). *Br. Polymer J.*, **4**, 27.
43. KIRK, P. G. and STARK, G. W. V. (1975). Flexible Polyurethane Foam: *Large Scale Fires of Industrial Loads of Seating Cushions*, Her Majesty's Stationery Office, London.
44. WILSON, W. J. (1976). *The Journal of Fire & Flammability*, **7**, 112, Jan.

45. *Flexible Polyurethane Foam, its Uses and Misuses*, The British Rubber Manufacturers Association, London, 1976.
46. PALMER, K. N. and TAYLOR, W. (1974). *Fire Hazards of Plastics in Furniture and Furnishings*, Building Research Establishment, UK Dept. of the Environment, Feb.
47. PALMER, K. N., TAYLOR, W. and PAUL, K. T. (1976). *Fire Hazards of Plastics in Furniture and Furnishings: Fires in Furnished Rooms*, Building Research Establishment, UK Dept. of the Environment, Feb.
48. CHANDLER, S. E. (1976). *Some Trends in Furniture Fires in Domestic Premises*, Building Research Establishment, UK Dept. of the Environment, Oct.
49. PALMER, K. N., TAYLOR, W. and PAUL, K. T. (1975). *Fire Hazards of Plastics in Furniture and Furnishings; Characteristics of the Burning*, Building Research Establishment, UK Dept. of the Environment, Jan.
50. DAMANT, G. H. and YOUNG, M. A. (1977). Smouldering characteristics of fabrics used as upholstered furniture covers, Lab. Report No. SP-77-1, State of California, Jan.
51. HANCOCK, E. (1977). Conference Report, *Cabinet Maker & Retail Furnisher*, 27 May, pp. 30–2.
52. BRISSON, M. Y. (1976). Int. Symposium, Toxicology and Physiology of Combustion Products, Salt Lake City, Utah, 22–26 Mar.
53. CARROLL, W. G. ICI Organics Division, Private communication.
54. KORZENIOWSKI, Z. and PIEKARSKI, K. (1975). *J. Cell Plast.*, **11**(1), 36–9.

Chapter 5

DEVELOPMENTS IN THE USE OF URETHANE POLYMERS IN THE TRANSPORT INDUSTRY

WALTER E. BECKER

*Mobay Chemical Corporation,
Pittsburgh, Pennsylvania, USA*

SUMMARY

Any in-depth discussion concerning the developments of polyurethane polymers in the transport industry would require a full book unto itself. This chapter, therefore, will concentrate only on four application areas in which significant changes are underway.

Flexible foam seating remains as one of the largest outlets for urethane polymers. Whereas this application was fairly static in terms of development during the mid to late 1960s, many major technological developments have been introduced since that time, including moulded foam seating, full foam seating, high resilient foam and combustion modified foams.

By far the most exciting technological development of polyurethane materials in the transport industry at the present time is that of reaction injection moulding (RIM). The necessity for damage-resistant front and rears for automobiles has led to a market opportunity for polyurethane materials anticipated to be of the order of £160 million per year by 1980.

Until the early 1970s, low density rigid urethane foam usage in the transport industry was confined to that as an insulation material in refrigerated railroad cars and trucks. Little, if any, use was made of other physical property attributes. The development of barge flotation technology and systems changed this situation dramatically. The outlook suggests that flotation uses for rigid urethane foam may well exceed the conventional transportation insulation requirements within 5–10 years.

*A relatively new application for polyurethane materials in the transport
industry is that of thermoplastic foam (TPI foam).* Polyurethane and
*polyisocyanurate technologies have been combined to create a family of
heat-formable foams and composites which may have a significant future as
automotive interior components, particularly head and door liners, sun
visors, etc.*

INTRODUCTION

The major applications for polyurethane materials in the transport
industry can be generally subdivided into four broad categories:

Flexible foam	—seating and trim
Elastomeric	—safety padding, bumpers, fascia and other exterior parts
Rigid foam	—thermal insulation, flotation
Coatings	—surface finishes for trucks, aircraft, boats, etc.

The US transport industry consumes over 400 million pounds of
urethane raw materials per year (Table 1).

TABLE 1

US TRANSPORTATION INDUSTRY CONSUMPTION OF URETHANE RAW
MATERIALS

	Millions of pounds				Forecast growth 1977–81
	1973	1974	1975	1976	(% per year)
Flexible foam	350	329	320	344	5–6
Elastomerics	8	10	10	12	50
Rigid foam	55	52	36	40	10

The abnormally high growth rate for elastomerics during the 1977–1981 era
is a function not only of the excellent growth prospects during this time but
a function as well of the present low consumption level which converts even
moderate absolute growth numbers into rather high percentage increases.

The consumption of polyurethane coatings raw materials for
transportation applications is more difficult to quantify because of the high
percentage of 'urethane modified' systems. It has been estimated, however,
that, in 1977, urethane coatings represented about 3 million gallons of the

total 65–70 million gallons of coatings materials consumed by the transport industry. A detailed treatise on the present and future application of polyurethane materials by the transport industry would require a textbook unto itself. Therefore, in this chapter, we will discuss in detail only those application areas where significant changes have recently occurred or are presently occurring. The areas to be discussed are:

1. Flexible foam seating.
2. Exterior components.
3. Barge flotation systems.
4. TPI foam (heat formable rigid foam).

FLEXIBLE FOAM SEATING

Polyether-based flexible urethane foams are widely used today in general transportation applications such as seating, internal trim, headliners, instrument panel padding, sun visors and carpet backing. The technology was developed by Farbenfabriken Bayer GmbH, Germany, and introduced into the US in the mid-1950s.[1,2] The first large volumes of commercial importance were made with the one-shot foaming technique and the slabstock process. About 1960 the hot moulding technique established itself and penetrated the market. In recent years a new generation of urethane foams, called high resilient foams, were developed. These foams are exceptionally resilient and elastic with a latex-like feel. They have become a major factor in seating because they can pass the MVSS 302 flame test[3] without the use of flame retardant additives* or with only a minimal amount required. (These values are based on laboratory tests and do not necessarily reflect an actual fire situation.)

In addition to excellent physical properties and durability, polyurethane foams offer an important contribution to energy savings. These energy savings come about due to reduced vehicle weight because full foam seat constructions can be produced at lower weight than spring-reinforced constructions. This lighter weight translates directly into improved miles per gallon. Furthermore, high resilient forms of polyurethane are essentially self-curing, significantly reducing the amount of process energy required to produce the finished seat.

* These products are currently described as combustion modifiers.

Production Techniques and Properties

Polyurethane foams are produced by reacting polyols containing an active hydrogen with isocyanate. Their physical properties are determined by the molecular structure of the resultant polymer. Accordingly, polyurethane foam products having a wide variety of properties may be prepared with the use of different building blocks in terms of polyol compounds, isocyanate compounds and selected additives. The word polyurethane is all inclusive and is used for all polymers containing urethane linkages, but it should be noted that the given polymers usually also contain other linkages. Essentially, there are seven major types of isocyanate reactions which are very important in polyurethane polymers; namely, urethane, urea, biuret, allophanate, aryl urea, dimer and trimer.[4,5]

The properties of urethane polymers will depend on the balance of these polymer types which in turn depends on factors such as the polyol type, isocyanate type, choice of catalysis, degree of cross-linking and processing conditions. Most of the mechanical properties will change with the molecular weight of the polymer as well as the molecular weight per cross-link or branch point. The urethane reaction is excellent for extending the polymer growth, whereas reactions such as the allophanate, biuret and aryl urea increase the molecular weight per cross-link in the polymer. With a good knowledge of the correlation between properties and polymer structure, a wide variety of urethane products can be produced and tailored to meet the requirements of many applications.

Catalysis is required to promote the reactions of isocyanate and compounds containing active hydrogen. The catalysts commonly employed today consist of very strong tertiary amines such as triethylenediamine and select metal catalysts such as stannous octoate. At low temperatures the tertiary amines promote reactions of isocyanate and water, and the metal catalysts promote the reaction of isocyanate and hydroxyl groups of the polyol compound. Although the role of catalysis in flexible polyether foam preparation is well known with tin–amine catalyst systems, the changes in foam technology during the last decade or so are considered to be largely attributable to changes in polyol and isocyanate structures. Flexible foams are usually made with 105–115% of the stoichiometric amount of isocyanate required for reactivity with the active hydrogens of the polyol, amines and water of the formulation and, most preferably, with about 110% of the stoichiometric amount. This small excess of isocyanate is employed in the formulation to (a) assure complete reaction of the reactants and (b) obtain higher load-bearing properties in the foam through cross-linking reactions during foam cure.

The production techniques used in transportation seating are as shown below.

Skived Cushions

In the mid-1960s, many automotive seat cushions and backs were fabricated from slabstock foam. All foam grades as slabstock were first produced on facilities such as the Hennecke UBT-65 high pressure foam machine using well-known one-shot foaming techniques. They were poured to a width of 80 in and to a height of 25 in to 40 in depending on the density. Rectangular-shaped pieces were cut out of the bun when the foam was at least 24 h old. These pieces were subsequently contoured into the shape of an automotive cushion with the use of cutting knives and abrasive buffers. After combining them with burlap backing, sail-cloth trim and adhesive, they were ready for installation in cars. Such skived cushions established themselves commercially and competed directly against hot moulded foam for a number of years. The participation of skived automotive backs and cushions came to a rather sudden and dramatic end when the automobile manufacturers shifted a major share of their seating construction to high resilient moulded foam.

Hot Moulded Foams

Hot moulded foam technology has been well covered in the literature.[6,7] The process steps are listed in Table 2. The moulding cycle for a cushion is short and simple. The mould, which has been preheated to a surface temperature of approximately 35 °C and coated on the inside surface with a mould release agent, is filled in a single pass under the stationary head of the foam machine. The conveyor speed and metering rates allow for the preselected amount of foaming ingredients to be charged into the mould. The mould is closed and immediately conveyed into the high capacity radiantly heated oven to increase the temperature at the mould–foam

TABLE 2
METHOD OF MOULDING A CELLULAR POLYURETHANE ARTICLE HAVING A POROUS
SURFACE

Step 1:	Inner surface of mould coated with a mould release agent
Step 2:	Mould filled with foamable polyurethane composition
Step 3:	Polyurethane foam formed
Step 4:	Mould heated to cure polyurethane and melt the mould release agent
Step 5:	Moulded article removed from mould while mould release agent is liquid

interface from 35 to 100 °C in 3–4 min to overcome the heat capacity property of the mould and enhance the foaming reaction. Thereafter, the mould is held for 6–7 min in a 175 °C convection oven to promote foam cure and to keep the wax release agent in a molten state. Thus, the entire mould cycle requires 9–11 min. When the cushion has been removed, the mould can be conducted through another moulding cycle. Each cushion is post-cured for 2 h in a 121 °C convection oven to optimise the physical properties.

High Resiliency Moulded Foam
High resiliency moulded foam is prepared with faster moulding cycles than employed for hot moulded pieces and with considerably lower mould temperature and little or no post-cure. Although the reaction rates are such that foaming can be done at ambient conditions, improved surface appearance on the moulded parts is obtained when the moulds are maintained at approximately 55 °C at pour time.[8] Much of the overall process technology is very similar to that of hot moulded foams, especially in terms of raw material handling, machinery and techniques. Early forms of high resilient moulded foams required crushing after demoulding to ensure open cells and to prevent shrinkage. Current developments are

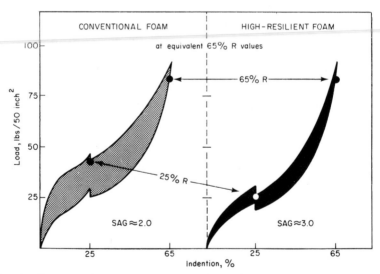

FIG. 1.   Stress–strain curves for conventional foam vs. high-resilient foam. R = rest value.

TABLE 3
TYPICAL STARTING FORMULATIONS FOR MOULDED AUTOMOTIVE
SEATS

Hot Moulded Foam

Formulation, parts by weight	
Polyether Multranol® E-9124[a]	100
Water	3·5
Niax A-1[b]	0·10–0·15
Silicone surfactant L 520[b]	1·0
Stannous octoate[c]	0·13
Mondur® TD–80®[a]	(~103 index)

High Resiliency Moulded Foam

Polyether Multranol® 3901[a]	60
Graft Polymer[a,b,d]	40
Water	2·9
Dabco[e]	0·3–0·6
Niax A-2[b]	0·1–0·2
Silicone copolymer L 5303[b]	1·0–1·5
Niax A-4[b] or DM-70[f]	0·3–0·5
Dibutyl tin dilaurate[b]	0·0–0·05
MDI/TDI blend (1:4)[a,g]	(100–108 index)

[a] Mobay Chemical Corp.
[b] Union Carbide Corp.
[c] Metal and Thermit Corp.
[d] Union Carbide Niax 34-28; Bayer PU 3119, PU 3699; Mobay Multranol® E-9148.
[e] Air Products and Chemicals Inc.
[f] Jefferson Chemical Co.
[g] Mondur® E-422; Mondur® E-446.

showing the way towards foam systems which will eliminate the need for crushing.

Typical starting formulations for the preparation of hot moulded foam and high resiliency moulded foam are shown in Table 3. Most of the hot moulded foams are made at densities near 2·0 pcf, whereas the high resiliency foams are generally moulded to a density near 3·0 pcf. Physical properties are very good with both foams. Some differences are noted in resiliency, elongation and hysteresis. The high resiliency foams exhibit significantly more resiliency, slightly lower elongation and less hysteresis than the conventional foams. The latter characteristic is illustrated in Fig. 1 which shows comparative stress–strain data on these foams. The indentation load deflection (ILD) was measured in terms of pounds per 50 square units at 25 % and 65 % deflections during application and removal of the load. As can be seen, the hysteresis loop is considerably greater in the conventional foams. It should be noted that this presents little or no difficulty in the normal application area for conventional foams. The SAC ($\equiv$ SAG) factor (65/25 ILD ratio) of the high resiliency foams is usually near 3, whereas the SAC factor of the conventional foams is usually near 2. Accordingly, a high resilient foam with ILD at about 25, for example, exhibits a very soft feel at the low deflection (25 %) and good support properties at the high deflection (65 %).

Combustibility Properties

Urethane foams cannot be made fire-proof in the sense in which ceramics are fire-proof. They (foams) are organic materials and will burn under the right conditions even if modified to include fire-retardant additives. However, their combustibility properties can be modified with additives or changes in polymer structure to the extent that, when tested by prevailing small-scale flame tests of low heat energy, they do not support combustion after removal of the source of ignition. A combustion-modified foam has a certain degree of resistance to ignition and burning as measured and defined by a given flame test. Entire texts have been written on the subject of polymer combustion. The best recommendation is to adopt a screening test which is recognised by the industry as indicative of fire-resistant performance under 'real fire conditions'. A few of these procedures are described in the following sections.

Automotive Interior Components

With the passage of MVSS 302, which went into effect with the 1973 model year, it became extremely important to develop flexible urethane foams for seating units which will meet this flammability specification. However, besides meeting the new safety standard, the automotive companies also required that the general level of physical properties of such flame-retardant foams was not to differ from the properties previously considered industry standards. Furthermore, it was required that the foams would remain flame retardant under various exposure conditions for at least twelve months. Technology exists to comply with this flammability standard with hot moulded foams using various halogenated phosphate ester additives.[9]

The new generation of urethane foams, called high resiliency foams, can pass several of the small-scale flame tests without the use of any phosphorous halogen containing additives when properly formulated. Others require a minimal amount.

Aircraft Applications

A critical area for urethane foam seat cushioning is in aircraft seating. One of the most stringent combustibility requirements presently specified for flexible foam is issued by the US Federal Aviation Agency for commercial aircraft or variation of this test as specified by the aircraft manufacturers. The test is FAA 25.853.

TABLE 4

FOAM RECIPE DESIGNED TO MEET THE REQUIRE-
MENTS OF FAA 25.853

Formulation, parts by weight	
Multranol® 7100[a]	100
Mondur® TD-80®[a]	42
Water	3
Additive E-9402[a]	2
Surfactant E-9920[a]	1·5
Stannous octoate	0·18
N-Ethylmorpholine	0·2
Mobay catalyst E-9400[a]	0·06
Phosgard 2XC20[b]	10
Isocyanate index	105
Physical properties	
Density, pcf	2·2
Tensile strength, psi	16
Elongation, %	200
Tear strength, pli	2·6

[a] Mobay Chemical Corporation
[b] Monsanto Corporation
These values are based on laboratory tests and do not necessarily reflect an actual fire situation.

As of late 1977, no definite standards were set by the FAA on smoke generation, but smoke densities of 100 after 90 s and a maximum of 200 after 4 min, as measured by the National Bureau of Standards smoke chamber, are presently contemplated.

Conventional and high resiliency foams can be formulated that comply with FAA 25.853 and exhibit low smoke generation (see Table 4).

ENERGY-ABSORBING EXTERIOR BODY COMPONENTS

Polyurethanes are already established as major contributors to automotive comfort and safety in the form of interior seating and padding. They made their first appearance on the exterior of the automobile in 1968 in the form of the 'Endura' bumpers on the Pontiacs. Endura bumpers serve primarily cosmetic and damage-resistant purposes rather than as energy-absorbing mechanisms. The appearance of these polyurethane bumpers, however, heralded the beginning of the era of energy management with polyurethane.

The rapid acceptance and growth of resilient bumper constructions was somewhat influenced by 1971 legislation which mandated safety and damage-resistant bumpers at impacts up to 5 mph. Of much greater influence, however, was the development of reaction injection moulding (RIM) technology.

The growth rate of elastomeric bumpers has been truly outstanding. Approximately 40 000 were produced for the 1968 models. By 1974, production was up to 800 000 units per year. Forecasts for the 1980 models call for the production in excess of 4·5 million units. In terms of pounds of polyurethane consumption, the 1980 forecast equates to approximately 160 million pounds.

The first production models were made from both cast microcellular polyurethane and thermoplastic polyurethane (TPU). The microcellular type soon became the material of choice, primarily because of lower manufacturing costs, often exceeding a reduction of 20 % as compared to the manufacturing costs of TPU. Some of the bumpers produced from microcellular urethane include:

1970–72	Pontiac Firebird
1971	Pontiac GTO
1973	Ford Mustang

The bumpers were rigidly mounted and had very thick sections, 1–3 in deep, at the leading edge.

Legislation requiring a 5 mph protection caused a shift in bumper design. The bumpers were moved outward from the car body and hydraulic shock absorbers were placed between the bumper and the car frame. The gaps were covered with filler panels or 'sight shields' produced of plastic or rubber.

The first attempt to incorporate the energy management function into the polyurethane bumper was that of the 1974 Ford Mustang. Here the sight shield was moulded as an integral part of the bumper, eliminating separate moulding and assembly costs. At present this bumper is still in production. The full front end concept had its first public showing on the 1973 Chevrolet Laguna. The polyurethane bumper in this case contained a metal insert and was mounted over hydraulic shock absorbers to meet the 5 mph requirement. The Chevrolet Laguna received 'The Product of the Year' award. This award is given to recognise innovation to automotive parts suppliers by *Autoproducts Magazine*.†

This first phase of elastomeric fascia development ended in approximately 1975. The technology at that time can best be summarised as follows. The components provided damage resistance from low mass impacts, such as stone and gravel, at very low speed and higher mass impacts at 1–2 mph. The benefits of scratch resistance and corrosion resistance were also provided. The components had thick sections which contributed to higher weight. The Laguna, for example, had 40 lb of urethane in the front end system. The total weight including hydraulics was over 100 lb. All of the parts contained steel inserts because of the low inherent stiffness of the polyurethane polymers. Production cycle times, which started at 15 min, were reduced to approximately 8–9 min by 1975.

The time span of 1973–1976 can be thought of as a transition phase. During this time, components were produced which demonstrated technology advances not present in previous cast urethane components. One of the first of these parts was the 1973 Pontiac Grand Am fascia. The programme started as a three-piece thin section (0·12–0·15 in thick) upper front fascia panel. Later it was converted to a one-piece construction. The programme was not fully cost effective, however, because it was based on a relatively high-priced thermoplastic urethane.

RIM flexible exterior components were first used in Germany in 1974 for the Ford Capri. Using polyurethane technology from Farbenfabriken Bayer GmbH, Ford developed a bumper which effectively was a C-shaped section, approximately 1/4–3/8 in thick, which was ribbed in order to absorb energy on impacts up to 5 mph. This design has been referred to as the 'Hi-Flex' design (Fig. 2).

The 1975 and 1976 Pontiac Firebirds had bumpers of cast microcellular urethane. Unlike earlier cast bumpers, however, polyurethane was selected to provide energy impact capability up to 5 mph, thus eliminating the need for hydraulic shock absorbers. The technological development this time

† *Autoproducts Magazine*, 21590 Greenfield Road, Oak Park, Michigan, 48237, USA.

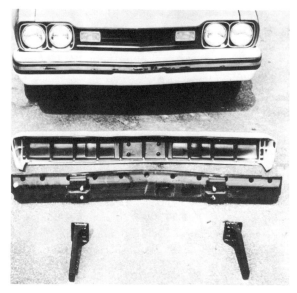

FIG. 2. Ford Capri Hi-Flex bumper.

had brought the microcellular foam density down to 40 pcf compared to the normal 50–65 pcf utilised during phase one. These components, however, represented cast elastomer technology and were not as fully cost and/or weight effective as subsequent RIM components.

The 1975 General Motors Chevrolet Monza 2 + 2 programme represented the introduction of RIM into commercial production. The programme included the Buick Skyhawk and the Oldsmobile Starfire as well as the Monza 2 + 2.

A two-piece construction was used for each front and rear end (total of four parts). The elements were thin sectioned at 0·150 in thick (12–16 lb of urethane for each two-piece end compared to 40 lb for the Laguna). This time, RIM technology had brought the cycle times down to 4 min and shortly thereafter 3 min.

The 1977 Firebird front fascia may well be the best representation of RIM technology as existed at that time. The front end was a one-piece unit which retained all of the benefits of previous multiple component constructions.

The 1977 Firebird front fascia used no hydraulics. All energy management for 5 mph impacts was effected by a low-weight urethane foam energy absorber.

Until this time, most of the applications for elastomeric urethane exterior components have called for materials with high impact resistance, particularly at low temperature but with relatively low stiffness (flexural modulus below 100 000 psi). The success of polyurethanes in these low modulus applications has generated a tremendous amount of interest and development activity in polyurethanes for non-fascia applications,

FIG. 3.   Impact test facility at Bayer Leverkusen.

including doors, bumpers, boot lids, etc. The requirements for these applications are slightly different in that they require high flexural moduli but may sacrifice something in low temperature impact. High modulus polyurethane materials (300 000 psi range) have already been developed which sacrifice very little in impact resistance. Further efforts in this area will undoubtedly generate more success.

Considerable interest and development activity have been initiated in filled polyurethane elastomers for applications requiring ultra-high flexural moduli (500 000 psi range).[28] This work is still in its infancy. However, it is expected that this will become a dominant area of RIM technology development and will open even broader horizons for polyurethane usage by the automobile industry. This technology may lead polyurethanes to a fully competitive position with other glass-reinforced plastics.

It is to be expected that continued progress will be made in the development of impact-resistant polyurethane materials. Test facilities, such as that shown in Fig. 3, will help to quantitatively predict the real world performance of polyurethane fascias. This unit, erected at the Leverkusen facilities of Farbenfabriken Bayer GmbH, is capable of impact testing fascias at room and low temperatures.

The history of RIM has been relatively short when compared with the overall polyurethane technological development. The first practical demonstration of RIM was the Bayer all-plastic car shown for the first time in 1967.

In terms of technological progress, however, the RIM development is a very important technological accomplishment by the polyurethane industry and has been well documented in the literature.[10-33]

A complementary market opportunity for polyurethane has been developed in conjunction with the RIM fascias opportunity, namely energy absorbing (EA) foam. The fascia *per se* provides very little in terms of energy absorption and/or energy management. While many theoretical concepts are possible for absorbing impact energy, primarily three methods are under consideration:

1. Hydraulic absorbers.
2. Metal or plastic honeycomb.
3. Energy-absorbing polyurethane foam.

It is too early in the development of these concepts to predict the ultimate winner. Polyurethanes can offer a considerable contribution to energy absorption at relatively low weight and cost and can eliminate the 'one time only' problem encountered with honeycomb inserts.

The final designs are obviously not available at the present time to make an accurate prediction, but it is quite probable that quantities of the order of 10–20 lb of EA foam per automobile might well be used in the future in conjunction with elastomeric fascias.

BARGE FOAMING

The first ten years of growth of rigid urethane foam in the transport industry can best be attributed to the outstanding thermal insulation capability of this product. Little, if any, use was made of the other attributes of rigid urethane foam. Certainly some credit was given for the lighter weight of the product with respect to other insulation materials, but weight itself was far from an important criterion in the selection of rigid urethane foam for use in transportation vehicles. This picture changed drastically with the development of urethane foam technology for barge flotation.

Urethane foam flotation has been utilised since the first days of the urethane foam industry. It has become the primary flotation material in the recreational boat industry, particularly where permanence of flotation was of major importance, but major use of rigid foam in flotation in larger vessels was quite limited.

In the early 1970s, the river barge industry in the US had a problem. Many of the barges in service were quite old, rather battered and very leaky. In that state they were basically unusable.

This problem of the barge industry was solved and a new growing market for rigid urethane foam flotation was created when it was found that barges could be rejuvenated routinely in dry-dock by filling the spaces around the cargo area with rigid urethane foam.

Because of its low density and low water absorption rate, urethane foam provides a buoyancy of 60 pcf. When all the voids (rakes, wing tanks and floors) are filled with rigid urethane foam, a barge is virtually leak-proof and sink-proof. A worn-out barge, thus rejuvenated, can go back into service at a fraction of the cost of a new barge.

Applications for barge flotation run the gamut from inland and ocean-going to deck, dredge hulls, booster barges and even pipe pontoons.

After nearly five years of practical experience, the following advantages can now be claimed for rigid urethane foam barge flotation.

1. Improved flotation—rigid urethane foam greatly reduces, if not eliminates the likelihood of sinking. All costs normally incurred with sinking (i.e. salvage costs, cargo costs and occasional total barge costs) are obviously reduced. The estimates indicated that, prior to 1975, more than 650 barges sank each year, of which only about 50 % were recovered.

2. Extension of barge life—filling adds at least five years to the life of a barge. Many rehabilitated barges at the age of eight years showed every indication of lasting at least an additional five. Construction costs for new barges have increased drastically, giving the fleet owners even more incentive to extend barge life. In addition, a rejuvenated barge is ready for service almost immediately, whereas a one to two year waiting period is incurred with a new barge.

3. Lower maintenance costs—normal barges require pumping out; at least every week and, in many cases, every day. A urethane foam filled barge requires no pumping. The number of times that dry-docking is necessary for routine maintenance is greatly reduced.

4. Greater structural strength—if the inner surfaces of the barge are

relatively clean and free of scale and water, excellent foam adhesion will be attained yielding a sandwich-type structure through which a significant structural strength improvement is imparted by the urethane foam. In many cases with older barges, it is difficult to obtain a scale-free, water-free surface, so the advantage of greater structural strength is not realised; however, in the construction of new barges, the greater structural strength attribute of rigid urethane foam can be a significant factor.

Large foaming facilities were installed in the early 1970s by several barge owners and contractors. One of these was Midwest Towing Company who erected a fast foaming operation at their dry-dock for barges near Ste. Genevieve, Missouri, on the Mississippi River.

The Midwest facility (Figs. 4–6) is capable of delivering 350–400 lb of foam system per minute, and completely foams a 195 ft long, 1500 ton cargo capacity barge in about 5 h.

Each shot that fills an inner bottom void space across the 35 ft width of a barge is precisely timed. The foam flows from the mixing head through an attached 20 ft long hollow aluminium tube called a probe, which is held by a crew of four workmen.

The shot begins with the probe fully inserted through the hole and to the void space as the foam shoots across and fills the opposite side of the space. Then, in intervals of 10–15 s, the probe is withdrawn in steps as foam fills the

FIG. 4.   Workers withdrawing 20 ft probe after shot.

FIG. 5.   Filling of wing tanks through hatches along the gunwales.

space smoothly and evenly. The crew generally moves on to fill the next adjoining space before the expanding foam oozes from one space to another through serrations in the transverse frames.

When foaming holes are cut along the centre line of the main deck for foaming a barge in the water with a conventional system, two shots are

FIG. 6.   Foaming machinery located in building 200 ft from the barge.

necessary to fill each void space; one shot for the starboard side and the other for the port side. Separate shots tend to result in gaps and air pockets in the foam. The single shot procedure employed at Midwest eliminates the problems with air pockets and gaps.

The key to consistent high quality of foaming is that only in dry dock can virtually all water be drained out of the inner bottom tanks. Drain holes are cut in with a torch in the bottom outer hull. Drainage also occurs through some 80 foaming holes in the side of the barge cut at 2 ft intervals along the lower edge of one side.

Most of the water that has leaked into the bottom tank readily spills out of these holes under the force of gravity prior to foaming. During the foaming operation, the expanding foam squeezes out the last traces of water. The result is that the foam has 90 % of its cells closed, so the foam will absorb essentially no water.

When a barge is foamed in the water with the foam insertion holes cut through the top of the tanks, there is no way for water to spill or be squeezed out of these holes. That is why, even with arduous pumping of tanks, usually one or two inches of water remain in the inner bottom to react with urethane chemicals and produce weak, spongy, moisture absorbent foam, which does not adhere to the tank's inner walls. Perhaps 50 % or more of the foam cells are open and interconnected, so they can absorb water. Only foam in the dry upper areas of the tanks is water-tight.

Foaming a barge in dry-dock, out of water, assures production of foam that is not affected by a special problem encountered when the water temperature is much lower or higher than the temperature of the air. When the barge is in the water, the lowest foam near the bottom outer hull is affected by the water temperature, while higher foam near the main deck is affected by the air temperature. This can result in foam that varies widely in quality between lower and upper areas of the inner bottom as temperature affects expansion of the foam. Foaming in dry-dock eliminates this problem.

Machinery that precisely meters and pumps the chemicals is in a small white frame building about 200 ft up the river bank from the dry-dock. The building is flanked by two huge urethane insulated tanks, each of which holds 6000 gallons or about 60 000 lb of one of the two major chemical components (an isocyanate or a polyol) each with certain additives. Liquid chemicals flow through 250 ft of hoses from the equipment building to the dry-dock where the specially designed mixing head combines the two components. They react instantly to form a foam which is shot into the barge.

The chemical components are propelled by two large positive proportioning piston displacement main pumps, each of which operates at 1800 rpm and can pump as much as 200 lb of chemicals per minute. Unlike gear pumps used in conventional systems, their operation is not affected by back pressures. To assure foam of consistent high quality, the heat exchanger maintains the chemicals at a constant 72 °F before and during foaming operations regardless of weather conditions. They can raise the temperature or lower the temperature of the chemicals as much as 8–10 °F/min.

The control panel for the entire operation is located in the equipment building as is the manifold through which chemicals pass from rigid pipes into flexible hoses. The lines are capable of being heated for winter foaming, even when the temperature is as low as 40 °F.

TPI FOAM—HEAT-FORMABLE POLYURETHANE RIGID FOAM

An interesting development with specific importance for the automotive industry has been the evolution of technology on heat-formable polyurethane rigid foams. It has been well recognised that urethanes have some thermoplastic nature and, therefore, are subject to heat deformation. For the most part, however, there has been no practical utilisation of this characteristic of the material. Traditional polyurethane compounds, if sufficiently thermoplastic to permit heat forming, did not provide finished parts with sufficient thermal dimensional stability to provide any useful function. On the other hand, materials which in their end form would have satisfactory thermal dimensional stability were normally so highly cross-linked that they presented difficulties in heat forming, if indeed they could be heat formed at all.

From the store of knowledge developed in the polyisocyanurate area came the necessary hints to permit the TPI foams to move from merely a concept into a totally viable product area.

A typical formulation for TPI foam is shown in Table 5. It differs from standard foam systems which incorporate a considerable number of branch points. In TPI foam, primarily linear reactions are utilised. The polyol component contains, along with the base polyol and catalyst, a small amount of water to ensure open cell character. Open cells greatly improve the heat formability.

One of the challenges is to develop an efficient process to prepare flat panels. One of the most efficient production processes for such products is the double belt laminator.[34] This equipment is well-known in the polyurethane industry by the manufacturer of laminates, primarily for the building industry. Only slight modification is necessary in order to produce TPI foam.

Liquid components are brought together in the usual manner and deposited via a mixhead onto a bottom substrate which travels under a

TABLE 5

FORMULATION FOR TPI FOAM

Polyol component
100·0 pbw polyether
0·5 pbw water
5·0 pbw catalyst
50·0 pbw blowing agent

Polyisocyanate component
400 pbw diisocyanate (MDI type)

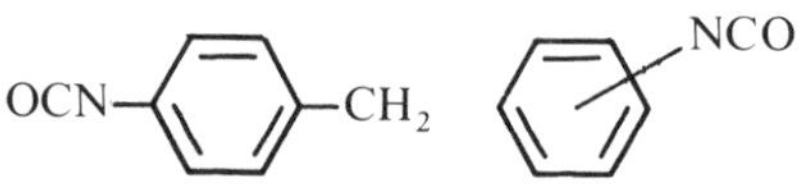

(MDI monomer type)

pbw = parts by weight

continuously moving conveyor, approximately 1–2 m wide by 12 m long (Fig. 7). A top substrate is introduced and, during entrance into the conveyor, the reaction mixture distributes outward and soaks into the upper and lower facings. This results in the development of edge zones of slightly higher density than the overall panel density. Shortly after entrance into the conveyor, the mixture begins to react, foam and expand. After curing in the conveyor, the finished panel is cut to length. Cure time in the conveyor is dependent upon machine output, panel thickness and conveyor speed. At present, it is expected that sandwich panels with a maximum 1·3 m width, 20–40 mm thickness and overall densities of 50–80 kg/m^3 will be desired.

Panels can be made with both surfaces of flexible foam or one flexible foam facing and an edge zone reinforced with non-woven fabric. Due to inventory problems, it appears, however, advisable to manufacture panels without a finished surface and apply this surface to the material during the forming step.

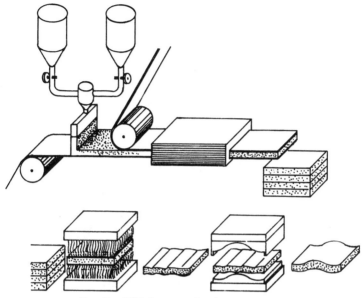

FIG. 7. TPI foam production process.

The moulding of TPI foam can be carried out in a semi-automated or completely automated installation.

Water concentration in the open cells can adversely affect the heating and deformation characteristics; therefore, the panels must be kept dry. They are brought from storage and fed into a single- or multi-stage heating installation where they are brought to a deformation temperature of approximately 200 °C and subsequently quickly fed into a press containing the die and mould. Maximum stamping pressures are of the order of 5 kp/cm^2 (Fig. 8).

Dwell time is dependent upon: part thickness, part density, part geometry, desired quality (thermal stability) and thermal conductivity in the mould.

For the production of a 20 mm thick, 50 kg/m^3 density panel to an overall part thickness of 10 mm requires approximately 40 s. After cooling in the mould, the part is removed and transferred to post-forming operations at another station.

Hot air ovens, infra-red heating units and contact hot plates can be used to heat the panels to the moulding temperatures. The heating process takes longer than the stamping process; therefore, multi-stage heating stations are recommended.

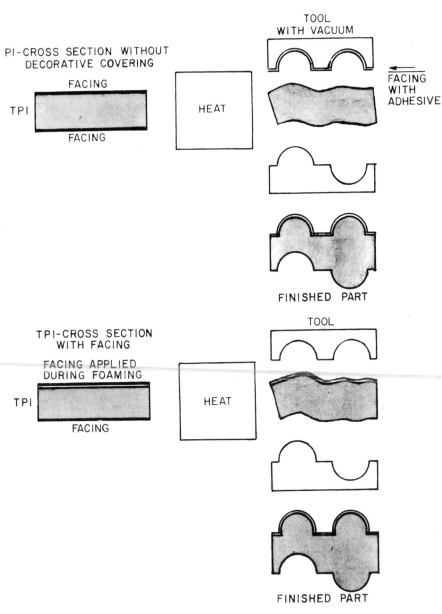

FIG. 8. Application of facings to TPI foam panels.

Epoxy tools have proven acceptable for prototype production. Wood tools in some cases have also been used. Because of the better thermal conductivity and shorter cooling time, metal tools, such as cast aluminium, are preferred for production. Moulds should be temperable and capable of accepting a vacuum on the finished side. They are built into quick-closing presses.

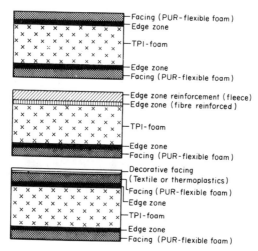

Facing (PUR-flexible foam)
Edge zone
TPI-foam
Edge zone
Facing (PUR-flexible foam)

Edge zone reinforcement (fleece)
Edge zone (fibre reinforced)
TPI-foam
Edge zone
Facing (PUR-flexible foam)

Decorative facing
(Textile or thermoplastics)
Facing (PUR-flexible foam)
Edge zone
TPI-foam
Edge zone
Facing (PUR-flexible foam)

FIG. 9.   Possible facing combinations for TPI foam panels.

Typical of the surface materials which are being used effectively with TPI systems are (Fig. 9):

(1)   thermoplastic sheets (PVC, ABS),
(2)   textiles (knitted and sewn articles),
(3)   flocking, and
(4)   polyester resin gel coat (long cure time).

Finished surfaces can be applied:

(a)   before stamping during production of panels (thermoplastic sheets, textiles with sufficient thermal resistance),
(b)   during the stamping (thermoplastic sheets, textiles, polyester gel coat), or
(c)   after stamping (thermoplastics, textiles and flocking).

The on-location, during-stamping process is preferred. The surface material is applied on the corresponding tool-half by vacuum, if possible, specifically for parts with deep or severe deformation. In extreme cases, the finished surface must first be thermally moulded. This can be done either on a separate tool or on the apparatus used for final part stamping. Thermoplastic and textile finished materials are furnished with a fusion-activated adhesive.

Finished parts can be fastened with melt screws, clips and similar attachments which can be applied during the stamping process. Adhesive bonding is also possible.

A primary target market for TPI foam has been the automotive market with emphasis on one-piece, snap-in headliners. Other areas under consideration are furniture, sports articles, orthopaedics, packaging, trailers and other automotive applications such as interior panelling for side doors and handles.

The response, thus far, of the European automobile industry to TPI foam has been very positive. Application processing is similar to existing metal processing techniques and the properties of the finished parts have been favourable. Parts produced from TPI foam are beginning to be utilised by the European automotive manufacturers.

REFERENCES

1. BAYER, O., MÜLLER, E. E., PETERSEN, S., PIEPENBRINK, H. F. and WINDEMUTH, E. (1950). *Rubber Chem. and Tech.*, **23**, 812.
2. HÖCHTLEN, A. (1952). *Kunstoffe*, **42**, 303.
3. US DEPARTMENT OF TRANSPORTATION, Federal Motor Vehicle Safety Standard 302, effective 1 Sept. 1972, distributed July 1977.
4. SAUNDERS, J. H. and SLOCOMBE, R. J. (1948). *Chem. Rev.*, **43**, 203.
5. SAUNDERS, J. H. (1959). 'Reactions of isocyanates and isocyanate derivatives at elevated temperatures', *Rubber Chem. and Tech.*, **32**(2), April–June.
6. US Patent 3,154,606, SZABAT, J. F. and MORECROFT, A. S. (1964). 'Moulding cellular polyurethane plastics', 27 Oct.
7. US Patent 3,210,448 SZABAT, J. F. (1965). 'Method of moulding a cellular polyurethane article having a porous surface', 5 Oct.
8. REID, G. E. and COFFEY, R. L. (1974). 'High resilient urethane foam, moulding production techniques and property variations', Second National Plastics in Furniture Conf., Atlanta, Georgia, 14 June.
9. SZABAT, J. F. and BAUMANN, G. F. (1974). 'Flammability aspects of flexible urethane foam', *Proc. of 1974 International Symp. on Flammability and Fire Retardants*, Canada.
10. PREPELKA, D. J. and METZGER, S. H. (1976). 'Advances in reaction injection moulding', *Advances in Urethane Science and Technology*, **4**, K. C. Frisch and S. L. Reegan, eds, Technomic Publishing.

11. LUDWICO, W. A. and CEKORIC, M. E. (Mobay Chemical Corp.) (1976). 'Automation of the RIM process', Paper 760336 presented at SAE Meeting, Detroit, Michigan, 23–27 Feb.
12. SCHAEFER, H. (1976). 'Advances in RIM tooling and technology', paper given at Purdue University, 8 June.
13. KNIPP, U. (1973). 'Plastics for automobile safety bumpers', *J. Cell. Plast.*, No. 2, March/April, 1–9.
14. ESTES, G. M., COOPER, S. L. and TOBOLSKY, A. V. (1970). *J. Macromol. Sci., Rev. Macromol. Chem.*, C. 4(1), 167.
15. KOUTSKY, J. A., HIEN, N. V. and COOPER, S. L. (1970). *J. Polym. Sci.*, Part B, 8, 353.
16. CLOUGH, S. B., SCHNEIDER, N. S. and KING, A. O. (1968). *J. Macromol. Sci., Phys.*, 2, C.41.
17. PEEBLES, L. H., JR (1974). *Macromol.*, 7(6), 872, Nov.–Dec.
18. HARRELL, L. L., JR (1969). *Macromol.*, 2, 607.
19. SCHOLLENBERGER, C. S. and DINBERGS, K. (1975). *J. Elast. Plast.*, 7, 65.
20. COOPER, S. L. and TOBOLSKY, A. V. (1966). *Textile Research J.*, 36, Sept.
21. CRITCHFIELD, F. E., KOLESK, J. V. and SEEFIELD, C. G., JR (1973). 'Thermoplastic polyurethane elastomers and temperature dependence of physical properties', paper presented at SAE Meeting, Detroit, Michigan, 14–18 May.
22. GERKIN, R. M. and CRITCHFIELD, F. E. (1974). 'Factors affecting high and low temperature performance in liquid reaction moulded urethane elastomer', paper presented at SAE Meeting, Toronto, Canada, 21–25 Oct.
23. PIGOTT, K. A., *et al.* (1960). *J. Chem. Eng. Data*, 5, 391.
24. SAUNDERS, J. H. and FRISCH, D. C. (1964). *Polyurethane—Chemistry and Technology*, Part II, 275–93, New York, Interscience.
25. SCHAEFER, H. (1974). 'Bayflex: a new material for elastomeric bumpers and body parts', paper given at SAE Meeting, Toronto, Canada, 21–25 Oct.
26. CARLETON, P. S., *et al.* (1974). 'RIM systems by computer techniques', paper given at SAE Meeting, Toronto, Canada, 21–25 Oct.
27. FERRARI, R. J., SILVERWOOD, H. A. and SALISBURY, W. C. (1976). *Plastics Technology*, 22(5), 39.
28. GUNNERSON, L. E. (1973). 'Shell Kraton G. thermoplastic rubber', paper given at SAE Meeting, Detroit, Michigan, 14–18 May.
29. LUDWICO, W. A. (1975). 'High modulus RIM elastomers for exterior automotive parts', paper given at SPI Meeting, Detroit, Michigan, 6–8 Oct.
30. ISHAM, A. B. (1975). Automotive fascias: how GR elastomers can make the grade, *Plastics Engineering*, Feb.
31. ISHAM, A. B. (1976). 'Glass fiber reinforced elastomers for automotive applications—a comparison of RIM urethanes and alternative material systems', paper given at SAE Meeting, Detroit, Michigan, 23–27 Feb.
32. PIECHOTA, H. and RÖHR, H. (1972). *Integralschaumstoffe*, Carl Hanser Verlag, Munich, 82.
33. *Autoproducts*, 'Plastics vs metals; which has the edge in the '80s?', June, 1975.
34. KRAFT, K. J. and BROCHHAGEN, F. K. (1966). 'The continuous manufacture of laminated insulating board from rigid polyurethane foam', *Kunstoffe im Bau*, No. 4, Strassenbauverlag, Heidelberg.

Chapter 6

DEVELOPMENTS IN THE USE OF URETHANE POLYMERS IN THE CONSTRUCTION INDUSTRY

D. J. DOHERTY and W. GREEN

ICI Organics Division, Manchester, UK

SUMMARY

Following a brief survey of the chemical raw materials used in the production of rigid urethane foam, the authors summarise the up-to-date information on the more important physical properties of the foam, viz. strength, dimensional stability, thermal conductivity and ageing. Particular attention is paid to the rate and mechanism of ageing of thermal conductivity.

It is shown that, from the data available, it appears that the very low values of thermal conductivity are most probably retained for decades at normal climatic temperatures.

The value of rigid foam as an insulant in energy saving is discussed. The various forms under which the foam may be manufactured and supplied are described. In conclusion specific instances of the use of rigid foam in the construction of cold stores, domestic housing and certain industrial and commercial buildings are described.

In 1965 only 10 000 tonnes of rigid polyurethane foam was used worldwide. By 1975 the total used in the construction industry had reached 180 000 tonnes of which the USA accounted for 60 000 tonnes and Western Europe about 75 000 tonnes. By 1977 the world figure for construction was in the region of 300 000 tonnes and the projected figure for 1978 is one-third of a million tonnes.

Rigid urethane and polyisocyanurate foam has become an important commodity in the construction industry. It has the lowest thermal conductivity of any insulating material on the market, and exhibits low water vapour transmission. It is light in weight with a high strength density relationship requiring a low level of structural support. The foam has good self-adhesive properties enabling it to be self-bonded to a variety of other materials, e.g. paper, metal, wood, plasterboard and bitumen felt. Furthermore, it is made from raw materials which are liquids which can be easily pumped and mixed permitting production *in situ* or in a factory. Finally, the foam-forming reaction is rapid and the final product is handleable within a few minutes of mixing.

The main application for rigid urethane and isocyanurate foam in the construction industry is for thermal insulation, although instances are increasing where the structural properties of the foam are of primary importance. Overall erection costs can be kept low since quick fixing techniques can be employed with factory produced composite structures. However, in some cases foam applied *in situ* is preferred.

CHEMICAL BACKGROUND

The chemistry of the formation of rigid urethane foam has been covered extensively elsewhere.[1,2] Throughout the industry the raw materials used are:

(a)   An isocyanate—usually polymeric MDI.
(b)   A resin which is normally an oxypropylated polyol (i.e. a polyether) derived from a limited number of feedstocks among which the better known are sucrose, glycerol, sorbitol, aliphatic and aromatic amines. Polyol mixtures are commonly used.
(c)   A blowing agent—normally Refrigerant 11, sometimes in admixture with water and/or Refrigerant 12 for 'pre-frothing', and/or water.*
(d)   Catalysts, surfactants, and fire retardants.

These materials can be supplied separately or compounded together (except for MDI) depending on the needs of the end user. Foam formulations can be adjusted to vary reactivity and flow for manufacturing purposes, and to confer specific physical properties on the end product.

The product most commonly manufactured is urethane foam (PUR) at a density of about 30–35 kg/m^3. More recently other polymeric species have appeared: isocyanurate (PIR), carbodiimide, polyimides and polyurea. Each of these products has polymeric MDI as a basic raw material.

* Refrigerant 11 = CFM 11. Refrigerant 12 – CFM 12.

Polymeric MDI

Total world manufacture of this product in 1975 was about 220 000 tonnes, 45 % of which was consumed in the construction industry. There are some differences between different manufacturers' specifications for the material, but most commercially available brands have a functionality in the region of $2 \cdot 8 \pm 0 \cdot 2$, an isocyanate value of 30–31 % and a viscosity at 25 °C within the range 1–4 poise. Hence, the specialised isocyanates based on refined MDI referred to in other chapters in the book have not been used to date to manufacture rigid foam for the building industry.

Polyols

There are a great number of these on the market, each with its own advantages and disadvantages. The choice of which one to use is dictated by the type of machinery being used or the properties of the end product, the processing characteristics of the polyol and the economics of the total manufacturing operation. It is not possible in a book of this nature to give guidance on which polyol or mixture of polyols should be chosen in particular circumstances—this normally should be done in consultation with the raw material suppliers or systems houses who are willing to offer advice.

Blowing Agents

Refrigerants 11 and 12 are pure chemicals sold for a variety of applications. There is very little technical difference between the products supplied by the manufacturers in this field, although some stabilised grades of Refrigerant 11 exist to retard ionic chlorine development. Typical stabilisers include α-methyl styrene and allo-ocimene.

Catalysts, Surface Active Agents and Fire Retardants

A wide variety of catalysts (amine and organo-metallic), surface active agents (mainly silicone based) and fire retardant agents (usually based on halogens and phosphorus) are available and are used in infinite permutations.

PHYSICAL PROPERTIES

Rigid foam is normally anisotropic because of the generation of flow lines in the foam during its formation. It is in this respect similar to timber and has an identifiable 'grain' direction. Some physical properties depend on the grain direction, which may vary from place to place in each article or composite. Other properties are independent of grain direction and may be regarded as bulk properties. The most important grain-dependent properties are strength characteristics and thermal conductivity. As for timber, the compression and tensile properties are greatest parallel to the grain direction, and least perpendicular to it. The aim of the manufacturer should be always to arrange his manufacturing processes in such a way that the optimum physical properties are obtained at the lowest cost. Thus in manufacturing a paper-faced laminate by the continuous process the aim is to produce a board having its maximum strength across the thickness so that in actual use as roof insulation the board withstands being walked on. This is achieved by allowing the foam to rise from the lower face vertically so that the maximum strength is in the vertical direction, and by ensuring that the grain is not subsequently distorted or forced laterally when the top facing is added.

Rigid urethane foam as normally used for insulation has a density of 30–$35\,kg/m^3$ ($\sim 2\,lb/ft^3$). Thus the product is about 97% gas which is contained in non-interconnecting cells with diameters in the range 0.2 to $1.0\,mm$. The physical properties therefore depend on the interaction and separate contributions of the gaseous and solid phases. Strength characteristics depend mainly on the solid phase, thermal conductivity depends on the gas phase and dimensional stability depends on both.

The sequence of events at the formation of a foam are extremely complex since a balance of forces has all the time to be maintained within a gas-in-liquid emulsion which is rapidly increasing in volume, temperature and viscosity simultaneously. The time scale is of the order of 1–2 min between the start and completion of foaming. At the end of this time the major volume expansion has ceased, the liquid phase has become solid, although increase in temperature may proceed for a time, until eventually the foam cools to room temperature. The gas in the cells is usually Refrigerant 11 ($CFCl_3$) with perhaps a little air and CO_2.

Strength

As already indicated, urethane and isocyanurate rigid foam are usually anisotropic in their strength characteristics. The anisotropy arises from the fact that the major cell diameters align themselves along the direction of flow of the material (grain direction) during the formation of the foam. In sealed cavities (i.e. by panel injection) it is possible to pressurise the foam so that the diameters are made more nearly equal and the anisotropy is reduced. Anisotropy is usually very small also for high density foams. Thus it is clear that the exact strength characteristics at any given density depend mainly on the method of manufacture and only to a secondary degree on the chemical formulation. However, with any given manufacturing process, consistency of strength characteristics can be maintained without difficulty. Strength is dependent on density. An empirical relationship has been found

$$s = kD^n$$

where s is the strength parameter, k is a constant, D is density and n is a power index close to 2.

In common with other plastic materials, rigid PUR and PIR foams exhibit viscoelastic properties. The materials are thermosetting and do not have easily measurable rubbery or liquid flow regions in the relaxation time spectrum. They do, however, exhibit load-bearing characteristics which are both temperature and time dependent. Formal viscoelastic characterisation of rigid foam species is extremely complex because of the interactions arising from anisotropy and changing cell gas pressures. *Ad hoc* methods of testing for strength characteristics are used, but their limitations should be clearly understood. The data normally quoted as the results of standard tests are compression, shear and tensile strengths. Bending or fracture strength may also be given. The data appertain to samples selected and tested according to standard procedures and as such they are used extensively for quality control purposes and for comparison of one product with another. For practical engineering purposes the figures may be used as a rough guide, but more extensive long-term testing is necessary to determine the structural properties of the foams with associated safety factors if these are indeed required. However, in the vast majority of current applications where low density foams are used as thermal insulants they are subjected to low levels of stress, which experience has shown they are capable of withstanding over many years without impairment of their properties.

Dransfield[3] has investigated the creep behaviour at 60 °C of rigid foams in compression. He showed that under constant load the strain in the material could be represented to a high degree of approximation by an equation of the form

$$E = E_0 + Kt^{1/6}$$

where E is the strain at time t, E_0 is the original strain and K is a constant. A typical curve illustrating his results is given in Fig. 1. This shows that even at 60 °C creep is extremely small, and at lower levels of loading and at lower temperatures creep should be significantly less. For example a compression strength of $2 \, kg/cm^3$ represents $20 \, tonnes/m^2$ and, allowing for a safety

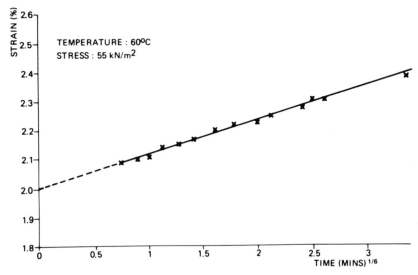

FIG. 1.   Creep behaviour of rigid urethane foam at 60 °C.

factor of 5, gives a useful long-term load-bearing capability of 4 tonnes/m^2 at room temperature. At low temperature the time dependency is very much reduced, so that at $-10\,°C$ (say) the safety factor could realistically be less than 5.

The stiffness characteristics of sandwich constructions has been studied in considerable detail for many years. They may be regarded as I-beams with a somewhat deformable web. Thus the deflection of a urethane-cored sandwich beam can be represented by:

$$\sigma = \frac{PL^3}{48D} + \frac{PL}{4GA}$$

where

$$D = \frac{E_f b}{12}(h^3 - c^3)$$

$$A = \frac{2hb\,(h^3 - c^3)}{3c\,(h^2 - c^2)}$$

and σ = central deflection of a simply supported beam, P = load at midpoint, L = span between supports, c = core thickness, E_f = tensile modulus of facings, b = width of beam, h = thickness of beam, G = shear modulus of core. The first term represents the classical Hookean deformation while the second shows the additional deformation due to shear in the core. The design and structural implications of this mode of behaviour were examined by Fisher[4] who recommended that, subject to certain limitations, urethane-cored panels were capable of sustaining the type of loads normally encountered in service. A more comprehensive treatment of this subject has been given by Hartsock.[5]

Dimensional Stability

Dimensional stability in the case of closed-cell foams is dependent on the ability of the foam to resist atmospheric pressure. If the internal cell pressure is too low, the foam may shrink unless the polymer network can withstand the inward pressure and, conversely, if the cell pressure is above atmospheric, swelling takes place unless the network can withstand these forces also.

When a foam is newly made the cell pressure is below atmospheric (Fig. 2), but on exposure to the atmosphere this can increase to a value above

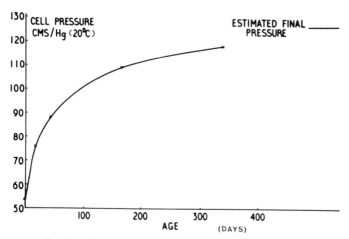

FIG. 2.   Changes in internal cell pressure on ageing.

atmospheric due to inward diffusion of air. The final value shown in Fig. 2 is due to the fact that the initial gas in the cells ($CFCl_3$) does not diffuse out quickly. The foam, to be stable, must resist the differential pressure. Properly manufactured foams will do this, but obviously as density is reduced the strength of the network is too low to resist the pressure differentials, and this is the factor which determines the lower density limits.

At present this figure can be as low as $27 \, \text{kg/m}^3$, although normally it is in the $32\text{--}35 \, \text{kg/m}^3$ region.

A complicating factor is the temperature dependency of strength. At $100\,°\text{C}$ urethane foam tends to lose a lot of its strength, whereas at lower temperatures it stiffens up. At the same time the internal gas pressure increases with rising temperature. Hence it is normal to carry out tests of dimensional stability over the range of temperature $-30\,°\text{C}$ to $100\,°\text{C}$. The measurements are sensitive to about 1 %. This is sufficient since instability, if present, manifests itself in massive volume changes which are easily visible.

Reversible changes in dimensions up to perhaps 0·5 % may occur as a result of temperature changes, or even ageing. These are largely due to changes in internal gas pressure, but the extent to which they may occur in any one sample depends on several factors. If the foam is bonded to rigid facings, then the only change detectable may be due to thermal movement of the facings, particularly if the thickness of the foam core is below 50 mm (say). If the sample is unfaced, or has only light-weight facings, some movement may be detected in the weakest direction, if the compression strength in this direction is in the region of $80 \, \text{kN/m}^2$ or less. At strength levels below this, complete instability may occur. It is, therefore, difficult to quote a classical coefficient of expansion for rigid foams, because the behaviour due to changes in temperature depends to a major extent on that of the attached facings.

Thermal Properties

Thermal conductivity (λ-value) is the most important physical characteristic of urethane rigid foam.[6-10] It has been shown that heat transfer is predominantly dependent on conduction through the gaseous phase, and the exceptionally low figure for rigid urethane foam is due to the presence in the cells of the fluorocarbon gas $CFCl_3$. A subsidiary factor is cell size, as transfer of heat by radiation is directly dependent on the cell diameter.

The conduction of heat in a foam may be described by the general equation

$$\lambda_f = \lambda_g + \lambda_s + \lambda_r + \lambda_c$$

where λ_f = thermal conductivity, λ_g = thermal conductivity of the cell gas, λ_s = thermal conductivity of the solid phase, λ_r = thermal conductivity associated with radiation across the cells, and λ_c = thermal conductivity due to convection within the cells.

It has been shown that $\lambda_c = 0$ for cell diameters below 10 mm, and λ_s is a constant at the normal foam density. λ_g and λ_r are the important variables

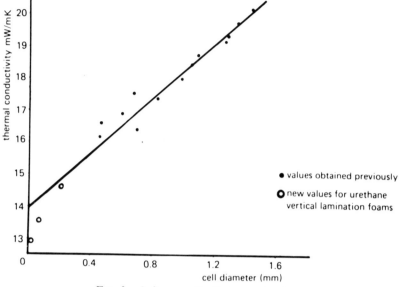

FIG. 3. Influence of cell size on λ-value.

which determine the performance of the foam as an insulant. It has been shown theoretically that the radiation component of conductivity should vary directly with cell diameter,[6] so that the larger the diameter the higher the conductivity. Buist et al.[7] showed in practice this was the case (Fig. 3). This means that other things being equal fine-celled foams will have lower conductivity than coarse grained ones. The exceptionally low λ-values associated with foams made by the vertical lamination technique are attributed to the exceptionally fine cells in the product.

λ_g is the parameter of most interest. In general it does not remain constant during the life of the foam because of diffusion of gases which may take place in and out of the cells through the cell walls. This is best illustrated by reference to the schematic diagram used by Ball[8] in explaining the effect (Fig. 4).

Point A (zero time) is when the foam is made and it contains mainly $CFCl_3$ in the cells with perhaps a little CO_2. The polymer is permeable to low molecular weight gases, so CO_2 diffuses out of the cells very rapidly and air starts to diffuse in. Soon the air starts to affect the conductivity (at B). From B to C there is a steady build-up of air in the cells until the partial pressure of air reaches barometric pressure at C. At this point the situation stabilises, because the rate at which the large $CFCl_3$ molecule can diffuse

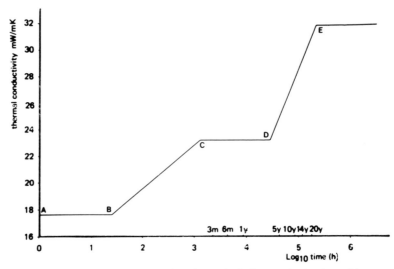

FIG. 4.   Schematic diagram showing theoretical changes in λ-value with age.

out is immensely low. No noticeable change in λ-value occurs until point D where loss of CFM11 through diffusion becomes sufficient to upset the existing thermal balance, and upward drift of λ-value occurs. This eventually reaches point E where the material has only air in the cells. If the foam is totally enclosed by non-permeable walls no shift in λ occurs, and the value remains permanently at A. However, it is difficult to guarantee complete exclusion of air, and it is usual to quote the λ-value as that at C. The rate at which level C is reached depends, therefore, on access to air, and surface to volume ratio of the product. For testing purposes, specimens of standard size (30 cm × 30 cm × 4 cm) are compared having been aged to equilibrium under standard conditions (23 °C, 50 % RH). Figure 5 is typical of the type of curve obtained. Typical λ-values are given in Table 1. These values are absolute, i.e. independent of dimensions or method of measurement.

Ball et al.[11] have carried out an extensive analysis of the variation in λ-value which occurs in practice. Both the difference between different

TABLE 1

	SI units	cgs	Imperial
Initial	0·017	0·015	0·12 Btu
Equilibrium	0·023	0·020	0·16 Btu

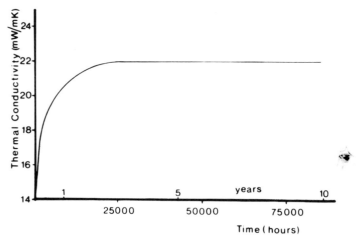

FIG. 5. Typical variation in λ-value with age.

manufacturing techniques and the manufacturing tolerances due to day-to-day variations within each of these techniques were examined. Figures 6 and 7 illustrate the statistical variation in λ-value for two of the processes considered—continuous lamination and slabstock. The main result of the analysis are given in Table 2.

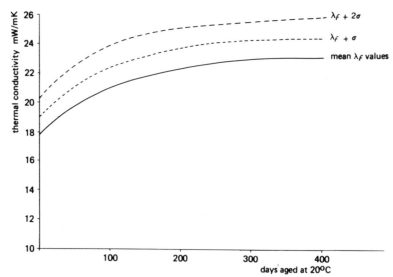

FIG. 6. λ-value ageing of laminate, showing the 95% confidence limits.

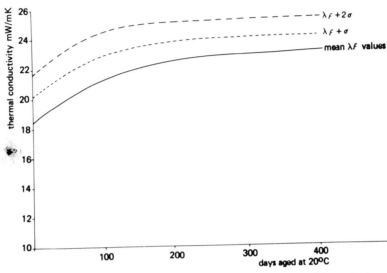

FIG. 7. λ-value ageing of slabstock, showing the 95% confidence limits.

The equilibrium values for laminate and slabstock are identical but somewhat higher than the values for press panels and vertical lamination. This difference is mainly accounted for by the radiative component of the conductivity. The scatter in the results represented by 2σ is remarkably

TABLE 2

Method of manufacture	Number of samples	Time in days	λ (mean value) (W/mK)	Standard deviation (σ) (W/mK)	2σ (W/mK)
Continuous lamination	39	0	0·018 5	0·001 6	0·003 2
	50	200	0·022 6	0·001 3	0·002 6
	27	400	0·023 2	0·001 0	0·002 0 (8·6%)
Slabstock	24	0	0·017 9	0·001 2	0·002 4
	56	200	0·022 4	0·001 4	0·002 8
	49	400	0·023 2	0·001 4	0·002 8 (12·1%)
Press injection	14	0	0·016 6	0·001 3	0·002 6
	78	200	0·019 9	0·001 3	0·002 6
	52	400	0·020 9	0·001 2	0·002 4 (11·5%)
Vertical lamination	19	0	0·014 9	0·001 0	0·002 0
	52	200	0·012 6	0·001 3	0·002 6
	11	400	0·018 5	0·001 2	0·002 4 (13%)

small, and is approximately constant, being independent of method of manufacture and the age of the sample.

Durability and Ageing

An important consideration for any product intended for use in the building industry is its ageing and durability characteristics. Rigid urethane foam is a relatively new product in this context, but information on its behaviour in service over a period of close to 20 years is now becoming available. In addition, the results of laboratory ageing studies can be used to supplement and confirm what is known from behaviour in the field.

From the earliest days it was known that rigid urethane had insufficient resistance to ultra-violet radiation. Thus it has never been used commercially in situations where it was exposed to direct sunlight. Laboratory tests also indicated that it had poor resistance to mineral acids, and moderate resistance to a wide range of organic solvents. The foam will withstand without difficulty both water and a wide range of petroleum products. However, immersion in such liquids over a period of years is to be discouraged particularly if liquid pressure is likely to be higher than a nominal few centimetres of mercury. Higher density foams or those having integral skins can be made which give a satisfactory performance for such specialist applications.

For the vast majority of applications, rigid foam is used in the form of a sandwich construction where the covering (or facing) is metal sheet, felt, plasterboard or any one of a range of flat or curved sheeting well known in the building and refrigeration industries. The temperature range over which these composites are used is approximately $-30\,°C$ to $+70\,°C$. There are specialist applications where foam may be used at temperatures as low as $-160\,°C$ and, for polyisocyanurate foams, up to $+140\,°C$.

In common with all polymers the rate of ageing of rigid urethane foam is dependent on temperature. Laboratory ageing studies have been conducted at a number of temperatures. In general terms these have established that the rate of ageing is not significant below $70\,°C$ for PUR. Some formulations are better than others in this respect. Some foam properties deteriorate at different rates from others. The subject, although well understood, is somewhat complex, and guidance from manufacturers should be sought.

Many detailed studies have been carried out to determine the effects of long-term laboratory ageing on λ-value. A detailed discussion of the latest information has been given by Ball et al.[11] Figure 8, taken from his paper, shows the results of 14 years laboratory ageing on a number of foams

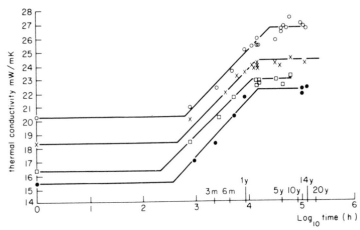

FIG. 8.   Laboratory test data at 20 °C, showing the effect of age on λ-value. Cell diameters: ○ = 1·5 mm, × = 1·1 mm, □ = 0·5 mm, ● = 0·5 mm.

differing in cell size but having the same density. There is no evidence of any drift from the equilibrium values. The samples were aged at 20 °C. Other samples with paper facings have been aged at 70 °C for 10 years and again no drift from equilibrium has been detected (Fig. 9). Under these circumstances it would seem that urethane foam at 20 °C should retain its λ-value for at least 50 years—probably much longer. With thicker sections and less access to air even greater safety is guaranteed.

It follows that with no change in the rate of outward diffusion of CFM11

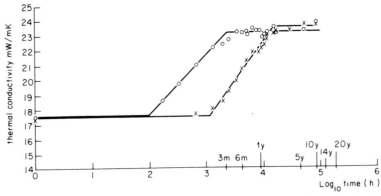

FIG. 9.   Laboratory test data at 70 °C, showing the effect of age on λ-value; ○ = ageing at 70 °C, × = ageing at 20 °C.

one may deduce that the polymeric species has not deteriorated in any way over this period. In particular the thin septa between cells have remained intact.

Evidence from actual performance in service supports the laboratory data. Indirect evidence can be quoted to the effect that the foam has now been in service in some applications for more than 15 years and to the best of the writers' knowledge no complaints of unsuitability or deterioration have occurred in that time, except in a few instances when the material was of inferior quality to start with.

Direct evidence is extremely difficult to obtain since testing of material in service involves the destruction of parts of installations which cannot be spared or which are costly to re-insulate. However, some work of this kind has been carried out in West Germany.[12] The work was carried out by the Munich-based Heat Insulation Research Institute.

Three separate test specimens were chosen from insulated roofs of two factories and a school. All three buildings had concrete decks and were insulated in the years 1965–66. The two factory buildings had each 30 mm of foam insulation and the school had 60 mm. Suitable test samples were removed from these roofs and tested for λ-value and moisture content. The results obtained are given in Table 3.

TABLE 3

Building	λ (at $10\,°C$)	Moisture content by volume	Years in service	Density
Factory 1	0·028 W/mK	0·008	$8\frac{1}{2}$	30
Factory 2	0·024 W/mK	0·063	$8\frac{3}{4}$	34
School	0·025 W/mK	0·34	$9\frac{3}{4}$	32

Hence after more than 8 years' exposure to the elements the foam samples retained their equilibrium λ-values, and remained impervious to moisture.

These quantitative measurements reinforce the mounting conviction of experts in the industry that rigid urethane foam is a long-lasting durable material. The product has now been in use in commercial quantities for 20 years and has withstood the test of time. Failures which have occurred have been due to the use of incorrectly manufactured material, or the exposure of the product to inhospitable conditions such as temperatures in excess of $100\,°C$, or too high fluid pressures. The experience gained over the years has given the industry confidence in recognising not only the limitations of the material, but also obvious advantages and longevity.

ENERGY SAVING CONSIDERATIONS

In recent years there has been increasing emphasis on the need to conserve the world's dwindling supplies of energy, and it is estimated that at least 50% of total world energy consumption is utilised for space heating purposes. The allocation of fuel uses within dwellings is shown in Figure 10.

In Britain, two-thirds of the energy used is consumed for space heating and fuel prices have risen dramatically during the 1970s, as shown in Table 4.

Different types of construction clearly influence the U-values for walls, roofs and windows but in an average semi-detached house the sources of heat loss are walls 35%, windows 17%, air change 19%, floor 9%, and roof 20% (Fig. 11). By the use of the proper combination of materials up to 25% reduction in fuel usage could be achieved for dwellings and savings could also be realised in commercial, industrial and agricultural buildings. In 1975 the British Rigid Urethane Foam Manufacturers Association

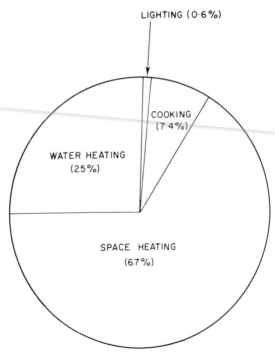

LIGHTING (0·6%)

COOKING (7·4%)

WATER HEATING (25%)

SPACE HEATING (67%)

FIG. 10.   Allocation of fuel uses within dwellings.[14]

TABLE 4
DOMESTIC ENERGY COSTS[13]

Year	Domestic heating oil (pence per gallon)	Gas (pence per therm)
1970	9·5	8·2
1972	11·4	9·3
1973	15·4	9·3
1974	23·0	10·0
1975	29·5	13·2
1976	37·3	15·1
1977	39·5	16·6

N.B. Gas price based on 1200 therms per annum.

(BRUFMA) recognised this fact and proposed insulation standards for all new UK buildings which would result in significant energy savings, shown in Table 5.

National governments throughout OECD are all taking steps both through financial inducements and by legislative means to save energy by the promotion of thermal insulation in domestic housing and commercial

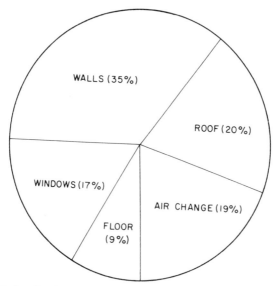

FIG. 11.   Relative distribution of heat loss of two-storey semi-detached house; gross floor area = 100 m², rate of conduction heat loss = 350 W/°C, ventilation heat loss = 80 W/°C, total loss = 430 W/°C.

TABLE 5

Number of years	Energy savings (million therms)
5	300
10	1 100
15	2 400
20	4 200
25	6 500 ($\equiv$ 14m tonnes oil)

premises. Table 6 summarises the situation for walls and roofs for eight European countries. Phase 1 refers to the practice prior to October 1973. Phase 2 gives the requirements 1976–77 and Phase 3 are proposed levels which may be implemented in the near future.

TABLE 6

Phases of national insulation levels described by transmission coefficients in $W/m^2 K$

	Phase	Denmark	France	Germany	Holland	Italy	Norway	Sweden	UK
	1	0·42	1·57	1·57	1·67	1·39	0·58	0·58	1·70
Walls	2	0·36	0·70	0·81	0·68	1·39	0·43	0·35	1·00
	3	0·30	0·41	0·47	0·57	0·36	0·27	0·30	0·55
	1	0·37	2·91	0·81	0·97	2·03	0·47	0·47	1·42
Roofs	2	0·27	0·55	0·69	0·68	2·03	0·33	0·25	0·60
	3	0·20	0·30	0·38	0·62	0·32	0·20	0·20	0·35

Source: Eurima News, December 1976.

It is estimated that on average 5 % of the total energy consumption of Western countries could be saved by the early 1980s by a reasonable level of insulation of existing buildings. Thereafter the savings increase as the stock of older buildings is replaced by newer ones where measures to conserve energy will include double or triple glazing, heat storage capacity, solar heating and a high level of thermal insulation. A number of studies have already been carried out on means of meeting these objectives.

Urethane rigid foam has a very important part to play in the campaign to save energy. As already stated, it has the lowest thermal conductivity of any commercially available insulant, it can be produced in a variety of ways and supplied in a number of forms and combinations to suit individual requirements. In addition, it is inert at normal ambient and sub-zero temperatures, it is unattractive to pests, fungi and insects and it retains its shape and dimensions for the duration of the life of the building.

Some of the studies which have been carried out show that a building can be insulated so well that no heat input is needed, other than that provided by electrical illumination to maintain adequate comfort conditions. A study was carried out on an industrial testing-house in the late 1960s which showed that no extra heat input was needed in winter, though some heat extraction was needed in summer. A recent experiment in domestic housing also showed that in specially designed houses energy consumption could be reduced to less than 10% of that used on average in the home.

Savings by insulating industrial premises are much more difficult to quantify, although they can be substantial. One practical application of urethane laminate to a factory roof at a cost of £4500 resulted in a 36% return on the capital invested in addition to much more pleasant working conditions.[15] A general study has been carried out of the effectiveness of insulating factory buildings in which it was shown that even modest application of foam insulation could reduce heat losses by up to 70%.[16]

PRODUCTION TECHNIQUES

Methods of production are diverse and can be summarised as follows.

(a) Continuous Lamination
Products ranging from 5 to 100 mm with a variety of rigid and flexible facings on one or two sides can be produced. Machines generally operate at 5–10 m/min but some machines can operate at speeds as high as 50 m/min. Product width is commonly 1·2 m and it is possible to reinforce the foam, e.g. with fibre-glass matt. The product is automatically cut at the end of the machine and stacked.

(b) Press Panel
This method is used to produce composite panels, invariably rigid faced for use as building elements. It is normal practice to inject 10–15% more foam into the panel than is required to just fill the cavity. This reduces the foam anisotropy and a minimum K-value is obtained. Because of the high pressure arising from the overpacking technique the panels have to be produced in a press, the occupation time varying depending upon panel thickness. Press occupation time is approximately 1 to 2 min/cm thickness. Output of the dispensing machines varies between 5 and 200 kg/min.

(c) *Vertical Lamination*

Permits the production of much larger double rigid faced composite panels than could be achieved by press injection. This is because of the much lower pressure generated during foaming by significantly reducing the degree of overpack. In this method the foam is deposited as a froth in a series of lifts.

(d) *Slabstock*

Both continuous and discontinuous slabstock techniques can be used for the production of large blocks of rigid foam. Blocks up to 1·5 m wide and 0·75 m high can be produced at rates of up to 5 m/min. Machinery similar to that used for flexible polyurethane foam blocks can be used.

(e) *Spraying*

This method has been used extensively in North America but only to a limited extent in Western Europe. It can be carried out in the factory, or on site, the foam being surface dry within seconds. American contractors have evaluated techniques for spraying roofs or walls, many acres in area, with uninterrupted foamed insulation which gives the maximum protection from ventilation or thermal bridging losses as well as providing a barrier to water penetration. Full protection of the foam is given by elastomeric membranes, metal sheeting or concrete.

APPLICATIONS

The range of applications for rigid foam in the construction industry is very large and it is not possible to deal with them all in detail. In the building sector it is easiest to classify the more important applications according to the functional use of the building; namely cold stores, houses and residential buildings, and industrial/commercial buildings.

Cold Stores

Cold stores and cold rooms are particularly suitable applications for rigid foam since the primary purpose of such buildings is to maintain a steady low temperature interior and they need, therefore, walls highly resistant to the passage of heat. Urethane foam, having the lowest λ-value of any commercially available insulant is clearly suitable for this duty. A large number of cold store panel manufacturers in Europe supply products for this purpose, using mainly a press panel technique. The panel thickness is usually within the range 150–200 m for stores operating below $-25\,^{\circ}C$. The length and width of panels on the market vary considerably. Many manufacturers offer packages which include interlocking panels designed to minimise heat leakage or 'weep' problems at the joints.

One of the earliest uses of rigid foam in cold store insulation was in 1959 in London. This store is still in operation without any problems. The more widespread use of rigid foam in this context did not get underway until the mid-1960s, and it has been growing ever since. A recent development in this context is the vertical lamination process.[17] Urethane foam made by this method is 10 % lower in λ-value than conventional urethane foam. This has led to the manufacture of panels 125 cm thick which have been found satisfactory in practice.[18] The better λ-value produced by this process mainly arises from the extremely fine texture of the foam. These panels with metal facings are 12 m long and have no divisions or other heat conducting fixtures anywhere along their length. Claimed U-values (in W/mK) for press and vertical lamination panels faced with 0·7 mm thick steel are:

Panel thickness (mm)	Press panel	Vertical lamination
75	0·244	
100	0·185	
125	0·148	0·114

A variation on this technique which is offered commercially is the use of glass-reinforced concrete facings 6 mm thick.[19] These facings are extremely effective in providing good mechanical strength and protection and also in offering a high degree of fire resistance.

Cold store panels can also be manufactured by a continuous lamination technique. The conventional method is to manufacture paper-faced slab and to glue rigid facings on subsequently. Recent machine developments would permit this to be done in one operation.

Houses and Residential Buildings

The manufacture of prefabricated houses from rigid foam sandwich constructions has been carried out in several parts of the world over the past 15 years. Several thousand houses of this type are in occupation—some for as many as 15 years—and as far as is known there have been no criticisms or complaints by the occupants on the grounds of structural strength or thermal performances. It should also be noted, perhaps, that there is no known record of fire damage or loss of life or property in these dwellings.

One of the earliest types of houses of this kind was erected in England in the early 1960s. These were one- and two-storey dwellings and both the outer walls and the internal partitions were made from urethane foam sandwich constructions. The normal design for exterior walls was an inner facing of plasterboard (1 cm), 75 mm of foam of nominal density $40 \, kg/m^3$ and an outer facing of 6 mm asbestos cement onto which an epoxy glue and decorative chippings were sprayed. The houses were designed to have a steel skeleton structure so that the panels were cladding panels only. Special care was taken to strengthen particular areas where sanitary fittings were attached to the wall. The panels themselves were made in a large $30 \times 8 \, ft$ metal press. The area was suitably divided up by timber so that multi-injections could be made. In all, some 2000 dwellings were manufactured in this way.

Another well publicised venture in house building utilises the concept of urethane foam containing fillers in the form of relatively large clay or glass balls sufficiently tightly packed for extra strength to be gained from the pressure of each ball upon the next. The foam at a density of about $70 \, kg/m^2$ acts both as an adhesive as well as a filler of the interstitial space. The balls are usually in the size range 10–30 mm diameter and the overall density is in the range 200–$400 \, kg/m^3$ depending on their composition.

The panels are manufactured in jigs by an injection technique whereby the end panel is first filled with balls, then the foam is injected through a plastic tube (or tubes) lying along the base of the panel with a number of holes along its length. A measured quantity of foam is injected which percolates into the spaces between the balls. The tube is not removed. Continuous methods are also used. The interior facing is normally plasterboard but the exterior facing is usually foamed against a release surface, and then conventional exterior rendering and finishing is applied. As in the former case, the houses have a steel skeleton structure and the panels act as external cladding.

One-storey houses and dormer-bungalows have been built from these panels. Most of the activity has been in France.[20] The advantages of using these panels are: low weight—thus reducing transport and erection costs; reduced heating costs; and lower expenditure on concrete and steel reinforcement. Sound insulation in the region of 30–35 dB is acceptable. From the builders point of view there is of course the advantages of products made consistently to a known specification, speed of erection because of prefabrication and consequently a fair degree of independence from the weather.

A new concept in house building has been devised in Norway.[21] A company who are experts in plasterboard manufacture have designed a plant to produce urethane-filled double plasterboard sandwich wall structures by continuous production methods. The board is unique because it simplifies building construction such that fully insulated houses can be erected in the form of a completely weatherproof shell within a few hours. Savings in time and cost of construction are expected to make the building elements increasingly attractive for large buildings as well as for houses, all insulated to the highest standards for comfort and economy.

Traditionally, plasterboard is mounted on site on a framework of wood or steel, with insulation sandwiched between the plasterboard layers. The new panels, however, provide the whole insulated wall structure in one piece from the factory. A special machine allows battens and sills, stored in large magazines, to be fed between two continuous layers of plasterboard and to be nailed in position. Electrical conduit is also automatically fed into place, and rigid polyurethane foam chemicals are dispensed between the two boards on a continuous basis. The product width is 2·4 m (the standard room height) and in principle any length of product can be produced. Currently the maximum length available is 8·4 m but ultimately 13·4 m will be possible. Consequently entire load-bearing external walls, partition walls and ceilings can all be pre-cut for a whole house and delivered on a single lorry. All the walls are 8 cm thick and contain 5·5 cm of urethane foam insulation. The basic structure, in the form of a weatherproof box with a lid, can be erected in 3 or 4 h. Apertures for doors and windows are cut out by bandsaw and window and door frames pressed into place. Electrical wiring is drawn through the built-in conduits and the roof trusses are placed in position and fastened to the battens incorporated during manufacture.

External walls are covered with weatherproof felt and covered with the customer's choice of material, e.g. wood, brick or marble pebble-dash, which serves as an external protection (Fig. 12).

The first test house was built in 1973 and in 1974 the Norwegian Ministry of Local Government approved the element for load-bearing external walls and partition walls. It has also been tested for fire resistance by Norwegian and Swedish authorities, and results showed that the external wall element with wooden cladding on battens meets the fire class B30 requirements for load-bearing structures. A large Belgian brick manufacturer also makes and markets polyurethane cored building panels.[22] The panels embody some striking technical innovations of proven worth for integrated 'instant' building systems. Their external facing of jointed clay briquettes, having the appearance of genuine brick are 19 × 5 cm in area and only 12 mm thick, and can be coloured red, yellow, brown or white. Jointing is by means of high density rigid polyurethane foam heavily loaded with sand. The inner

FIG. 12.   Norwegian house constructed from urethane foam-cored sandwich panel.

facing of the panel is a 6 mm thick waterproof fire-resistant board, and the edges are made of rigid polyurethane foam reinforced with glass fibre, grooved or equipped with projections as necessary for the constructional work. The inner core of the panels is of rigid foamed polyurethane ensuring good insulation properties.

The first prototype house was built at Braine le Chateau near Brussels in 1973. Since then, fuel savings of up to 50% have been recorded and the durability and weathering characteristics of the system have proved outstandingly good (Fig. 13). One of the largest Belgian construction companies chose these panels for some 1700 houses that were required in various Middle East countries where keeping heat out is as important as heat retention in many other countries.[23]

Clearly cost is of paramount importance and in mid-1977 a house kit based on such panels was selling at approximately 5400 Belgian francs per square metre of floor area, f.o.b. Antwerp. It is believed that such a dwelling could be erected in the Middle East for 12 000 Belgian francs per square metre floor area, excluding cost of land. Erection takes three men one week, excluding the foundations.

A great deal of thought on prefabricated housing in the Middle East has been generated since 1973. A multitude of architects and contractors are actively studying the problems involved. Local and foreign contractors are offering sandwich panels of various kinds. The experience gained in this area should be invaluable in establishing more firmly in the future the principles and advantages of prefabrication.

Prefabrication, however, is not the only way in which rigid urethane and isocyanurate foam finds application in buildings. Urethane foam laminated

FIG. 13.   Belgian house constructed from urethane foam-cored sandwich panel.

to plasterboard has been used to cure condensation problems arising from lack of ventilation. A similar product can be used to provide internal wall insulation where houses do not have a cavity wall. One enterprising foam producer utilises chopped waste foam to fill wall cavities instead of urea formaldehyde foam.

The facts that rigid foam laminated to flaxboard is used for the construction and insulation of roofs of new houses in Holland whilst foam laminated to plywood is used for the floors of caravans illustrates the wide range of applications.

In Britain *in situ* application of rigid urethane foam has solved a serious problem which occurs with existing houses.[24] In cavity wall construction, metal ties, bedded in the mortar courses, are used to hold the two leaves of brick together. Occasionally problems arise because the ties corrode due to high chlorine or sulphur contents of the mortar or poor metal quality. Gradual corrosion results in a laminating effect on the metal which swells, forcing the brickwork apart leading to cracks appearing. It is usually the external leaf of the wall that shows the signs of failure. Subsidence can produce a similar effect. Previously the only solution was to demolish the outer leaf, replace the metal ties and rebuild the leaf. In 1975 for ordinary semi-detached houses the cost would be £4000 for 100 m^2 of wall and would take about 8 weeks. Now the problem can be rectified by injecting a special grade of urethane foam in the cavity. This glues the two leaves together at a very much lower cost, takes less than two days for an average house and gives the added bonus of improved insulation. The same technique has also been employed on flats and public buildings to stabilise walls and also cure problems of water ingress. Other insulants either cannot be applied *in situ* or do not have sufficient strength.

Industrial and Commercial Buildings
A great deal of thought has already been devoted to studies on effective methods of insulating different types of building. The separate contributions of roofs, floors, windows, walls and ventilation have been measured and studied, but in the main it has been shown that the main areas of heat loss are the walls and the roof. There are the areas where the use of rigid urethane foam can have maximum impact.

The use of urethane for insulating roofs is now well developed (see Fig. 14). It is normally used in the form of laminated boards which are laid using asphalt or bitumen in the conventional way on flat concrete roofs or steel decks. Special boards made from temperature-resistant isocyanurate foam are available for situations where asphalt is used as the adhesive. Specifications for such boards and literature describing application techniques are available from the manufacturers and trade associations. Thickness in Northern Europe ranges between 25 and 60 mm. Large areas of roof in excess of 100 000 square metres have been insulated with urethane foam in this manner, and savings as high as 40 % of the heating bill have been achieved by the use of urethane roofing board in old factories.

An alternative method of insulating both old and new roofs is by the use of sprayed foam. The liquid chemicals are sprayed directly onto the substrate over its whole area so that a continuous water impervious envelope is produced. This technique is used extensively in North America,[25,26] but to a very much lesser extent in Europe.[27] The foam must be protected from solar radiation and from danger from fire. Flexible sheeting is normally used for this purpose, although spray-on or roll-on coatings which offer substantial protection are also available.

Composite panels consisting of profiled steel facings with rigid urethane or isocyanurate foam are being increasingly adopted in the building

Fig. 14.   Roof of the National Exhibition Centre, Birmingham, insulated by H. H. Robertson Ltd with rigid polyurethane foam produced by Coolag Ltd. Reproduced by courtesy of Coolag Ltd, Glossop, Derbyshire, UK.

industry for both roofing and wall applications.[28,29] These are factory made products which are therefore of a reproducible and consistent quality and which have the considerable advantage that erection costs are much lower, because dependence on weather is less and total fixing time is significantly reduced over that needed to insulate a roof *in situ.*

For walls, insulating linings are available which often consist of single rigid facings of steel, plasterboard or asbestos board, attached to a rigid urethane core of any desired thickness in the range 15–50 mm, and a flexible facing which can be paper, aluminium foil, glass fleece, etc. The choice of facings will depend on the proposed end use of the materials where such considerations as fire properties, resistance to impact, cleanability, vapour permeability, etc. have to be taken into account.

Other Construction Outlets

Outside the normal building field there are numerous areas where urethane foam is used, primarily, but not always, for insulation purposes. Pipe insulation is an area which has shown increasing potential, but slabstock, moulded sections, pipe in pipe (insulation between two pipes), spraying and continuous spiral production techniques can be used to produce suitable insulation. Insulation of one pipeline, 55 miles long, has resulted in a saving of 4500 gallons of heating oil every day, equating to a saving of £0·65m per annum at 1977 fuel oil prices. On a larger scale, Owens-Corning Fiberglas Corporation were awarded a contract, in 1974, for the insulation of 389 miles of Alaska pipeline.[30] The insulation used was a mixture of fibre-glass and rigid polyurethane foam which had to be physically strong enough to endure the weight of snow and ice and the stress caused by thermal expansion and winds. Crude oil enters the pipeline at temperatures as high as 50 °C. At full capacity, oil flows at over 7 miles per hour and when full the line contains 9 million barrels. It is necessary for the thermal insulation to maintain a minimum temperature of -30 °C for the steel pipe in the event of a 21 day winter shutdown.

Refrigerated transport and insulation of storage tanks and ships also consume considerable quantities of urethane foam. In LNG tankers it is necessary for the foam to withstand temperatures as low as -160 °C. Present constructions involve placing up to five metal tanks within the ship hull but separated from it by insulation. As well as acting as a thermal insulant it must act as a secondary barrier in the case of fracture of the tanks, preserving the hull from contact with the fuel. The ability to apply sprayed urethane rigid foam *in situ* is thus a major advantage in obtaining a leak-free insulation barrier, being superior to the use of balsa wood, perlite and foamed PVC in this respect.

These are a few of the ways in which rigid urethane foam has been used and adapted to solve a number of problems. No doubt the future will reveal even more outlets for this most versatile of all insulants.

REFERENCES

1. SAUNDERS, J. H. and FRISCH, K. C. (1962). *Polyurethanes: Chemistry and Technology*, Part I, Wiley.
2. BUIST, J. M. and GUDGEON, H. (Eds.) (1968). *Advances in Polyurethane Technology*, Maclaren.
3. DRANSFIELD, A., ICI Organics Division, Private communication.
4. FISHER, B. H. (1965). *Trans. Plast. Inst.*, Lond., Supplement No. 1267.
5. HARTSOCK, J. A. (1968). *Design of Foam Filled Structures*, Technomic Publishing Co., Stamford, Conn.
6. DOHERTY, D. J., HURD, R. and LESTER, G. R. (1962). *Chemistry and Industry*, **30**, 1340.

7. BUIST, J. M., DOHERTY, D. J. and HURD, R. (1965). *Progress in Refrigeration Science and Technology*, 271.
8. BALL, G. W., HURD, R. and WALKER, M. G. (1970). *J. Cell. Plast.*, **6**(2), 66.
9. SCHMIDT, W. (1968). *Appl. Phys.*, **11**(4), 19.
10. NORTON, F. J. (1967). *J. Cell. Plast.*, **3**(1), 23.
11. BALL, G. W., HEALEY, W. G. and PARTINGTON, J. B. (1978). *J. Cell Plast.*, in press.
12. ZEHENDNER, H. Forschungsinstitut für Warmeschutz e.V., Private communication, Aug. 1975.
13. *Digest of UK Energy Statistics* 1977. Publication by HMSO from Department of Energy.
14. HUMPHREYS, K. F. J. (1974). The case for improved standards of thermal insulation, paper given at QMC, University of London, 19 Sept.
15. *Works Management*, Oct. 1976, pp. 71–2.
16. *Chartered Mechanical Engineer*, Nov. 1976, pp. 72–5.
17. MAYRHOFER, P. (1975). Polyisocyanurate foams, Paper given at 5th International Foam Symposium in Dusseldorf, May.
18. Hemsec literature published by C. Hemmings & Co. Ltd, Prescot, Merseyside, England.
19. Literature published by Veldhoen–Isolatie BV, Raalte, Holland.
20. *Modern Plastics International*, Mar. 1971, pp. 14–16.
21. *Gipsotex Multi-Element*, literature published by Den Norske Gipsplatefabrikk, Svelvik, Norway.
22. Literature published by Mosabrik SA, Lanklaar, Belgium.
23. Information from Ets. Fancois, Brussels, Belgium.
24. *Tyfoam* literature published by Urethane Foam Operatives & Co. Ltd, Llanelli, Dyfed, Wales.
25. *Roofing Siding Insulation*, June 1976, pp. 58–9.
26. *Roofing Siding Insulation*, June 1977, pp. 68–9.
27. *Plastics & Rubber Weekly*, 14 Jan. 1977, p. 19.
28. *Modern Plastics International*, Mar. 1971, p. 17.
29. Literature published by H. H. Robertson Ltd, Ellesmere Port, Wirral, England.
30. *Urethane Industry Digest*, **14**, Nov. 1974, pp. 167–9.

Chapter 7

IMPROVEMENTS IN THE FIRE PERFORMANCE OF RIGID POLYURETHANE FOAM

R. Hurd

ICI Organics Division, Manchester, UK

SUMMARY

Modifications to the chemistry of polyurethanes and other isocyanate based foams which have led to improvements in fire resistance are summarised. In particular, reference is made to the development of isocyanurate and carbodiimide foams; also to polyols based on trichlorbutene oxide and on epichlorhydrin chemistry.

It is emphasised, however, that the performance of the rigid foam in actual fire situations is greatly dependent upon the composite design. Data are presented illustrating this point and it is suggested that advances have been made in understanding composite performance. The results of some large-scale fire trials evaluating the performance of components are summarised.

The growing importance attached to the evaluation of smoke which hinders visibility and/or presents a toxic hazard is discussed. The results of some of the work comparing the toxicity of smoke from natural and synthetic materials are reported.

The fact that the benefits obtained in various applications by the use of rigid foams needs to be taken into account when assessing any hazard is emphasised.

INTRODUCTION

Fire statistics are still not sufficiently comprehensive to allow conclusions to be drawn on the effect of an increased usage of plastics in general and urethane foams in particular on fire hazards. The Flammability Research Center at the University of Utah have published statistics[1] which are extremely detailed in their breakdown of the causes of fires. They examine the number of cigarette-related fires, gasolene-related fires, etc. and these, together with other statistics emerging mainly in the USA, should enable the contribution made by plastics to fire hazards to be more readily identified.

Einhorn[1] has also pointed out that the proportion of deaths in the USA from fire has not changed in the period from the mid-1920s to 1974, and that since the 1920s was a pre-plastic era this would indicate that the introduction of plastics has not had a significant effect. The deaths from all types of fires in the USA were 62 per million in 1966 and 57 per million in 1972. The usage *per capita* of urethane foam increased approximately three-fold in the period 1966 to 1972 so there would again appear to be no cause to argue that the increasing use of urethane foams has led to a higher number of fire deaths.

The following extract from a paper by Stark[2] is also relevant to this theme:

An examination of statistics of casualties in fires, extracted from fire reports collected by FRS on behalf of the Home Office, for the period 1955 to 1972, has indicated that deaths attributed to fire gases and smoke now account for about half the total but that over the period examined, during which there had been a dramatic increase in the usage of plastics in building, the proportion of deaths to total casualties has remained the same. It can therefore be inferred that the introduction of plastics into dwellings has not introduced an increased risk of death of fire casualties. However, as the total number of casualties over the period has shown a three-fold increase, it can be inferred that changes in building design and their contents have encouraged the rapid development of fire and have made it more difficult for occupants to escape.

In the above paper, Stark also states that there is no evidence of synergism between carbon monoxide and hydrogen cyanide which are evolved when urethane foams are burned.

Many of the more irresponsible attacks on urethane foams are based on claims that lethal levels of HCN develop quickly or that 'super toxicants' are evolved uniquely from urethanes. Advances in knowledge such as those in the first two references are important since, as criteria, tests and statistics become available, they help a proper assessment of rigid foams to be made. It is clearly recognised by scientists working in this field that there are many difficulties in allocating responsibility for increased fire hazards, if any, over the years. The demand for increased comfort has increased thermal and acoustic insulation. Chimneys have been removed in many homes and draughts eliminated. Temperatures are higher, humidity lower and improved standards of living have led to an increased fire load in the home. It is, of course, sensible to attempt to minimise the conflict between safety and comfort so that safety can be maintained while comfort is achieved. With this objective the urethane industry has spent a great deal of resources over many years to improve the fire performance of rigid and flexible urethane foams and to increase the understanding of the factors affecting fire performance.

THE RELATIVE IMPORTANCE OF FIRE HAZARDS

Before attempting an assessment of the improvements which have been achieved in the fire performance of rigid foams it is important to identify the relative importance of hazards arising in fires. Although still under debate, the following assessment of the relative importance of the hazards in real-life fires (descending order of importance) probably represent the majority view at the present time:

Reduction of oxygen and increase in carbon monoxide;
Development of high temperature;
Smoke;
Direct consumption by fire;
Presence of toxic gases other than carbon monoxide; and
Development of fear.

THE EFFECT OF CHANGES IN THE CHEMISTRY OF RIGID FOAMS ON THEIR FIRE PERFORMANCE

In assessing what improvements relevant to the hazards described above can be made by chemical changes to rigid foams, the following are some of the properties to be examined:

(a)   ease of ignition (ignitability);
(b)   speed of flame spread;

(c)   heat evolution (release);
(d)   smoke;
(e)   toxic gases; and
(f)   ease of extinction.

A simpler classification under the following headings is sometimes used.

Combustibility;
Ignitability;
Reaction to fire; and
Resistance to fire.

As will be argued later, however, it is important to determine whether it is relevant to test the foam alone or whether a more meaningful result will be obtained if the composite to be used in the end application is tested.

IMPROVED FIRE RESISTANCE BY MODIFICATIONS TO FOAM CHEMISTRY

Effect of Isocyanate Type

Among the first and early improvements made to the fire performance of rigid foams was the development of diphenyl methane diisocyanate (MDI) to replace toluene diisocyanate as the preferred isocyanate for applications in the building industry. This was pioneered in the UK and has spread world-wide.

The MDI foams had a higher softening point, greater resistance to low temperature ignition and when burning did not melt and drip. Differences in the properties of MDI and TDI rigid foams were reported by Ball et al.[3]

Further improvements resulted from the introduction of polyether polyols based on aromatic and cyclic initiators such as toluene diamine, diamino diphenyl methane and sucrose and this further enhanced the char-forming properties of the derived foams.

Effect of Fire Retardant Additives

Fire retardants also became increasingly important and they tend to fall into two classes: additive and reactive. Examples are:

(i) *Additive*

$$O=P \begin{cases} OCH_2\text{—}CH_2Cl \\ OCH_2\text{—}CH_2Cl \\ OCH_2\text{—}CH_2Cl \end{cases}$$

trichloroethyl phosphate

(ii) *Reactive*

$$O=P \begin{cases} O[CH_2(CH_3)CHO]H \\ O[CH_2(CH_3)CHO]H \\ O[CH_2(CH_3)CHO]H \end{cases}$$

oxypropylated phosphoric acid

Phosphorus/halogen additives tend to reduce the ignitability and surface spread of flame of naked foam but can lead to an increase in smoke development when combustion occurs.

Ball *et al.*[4] have reported the effect in a number of small-scale fire tests of phosphorus flame retardants on several types of rigid foam.

Birky, Einhorn *et al.*[5] have reported improvements to flame penetration ignition and flame propagation by the use of additives but noted that retarded polymers were generally worse from smoke hazard potential. They also reported that tris 2,3 di-bromopropyl phosphate was effective by gas phase inhibition, whereas the other retardants tested were effective inhibitors in the condensed phase.

More recently, further data on the effect of fire retardants have been discussed by Einhorn *et al.*[6] and by Weil and Aaronson.[7]

The latter authors criticise those writers who have elevated a few observations to a rule which states that flame retardants always elevate the CO and HCN yield. Further, they produce data showing that this generalisation is incorrect.

Polyisocyanurate Foams

Ball *et al.*[3] pointed out in 1968 that further improvements in cross-link density within the context of conventional rigid urethane polyol technology would be difficult and argued that better results would be obtained by the incorporation into rigid urethane foams of specific ring structures of known inherent stability. These authors referred to earlier work on the cyclisation of three isocyanate groups to form an isocyanurate ring and described development work on the practical exploitation of the trimerisation reaction of MDI leading to the production of technically useful rigid isocyanurate foam (PIR foam) materials.

The cyclisation of MDI to form an isocyanurate ring can be represented as follows:

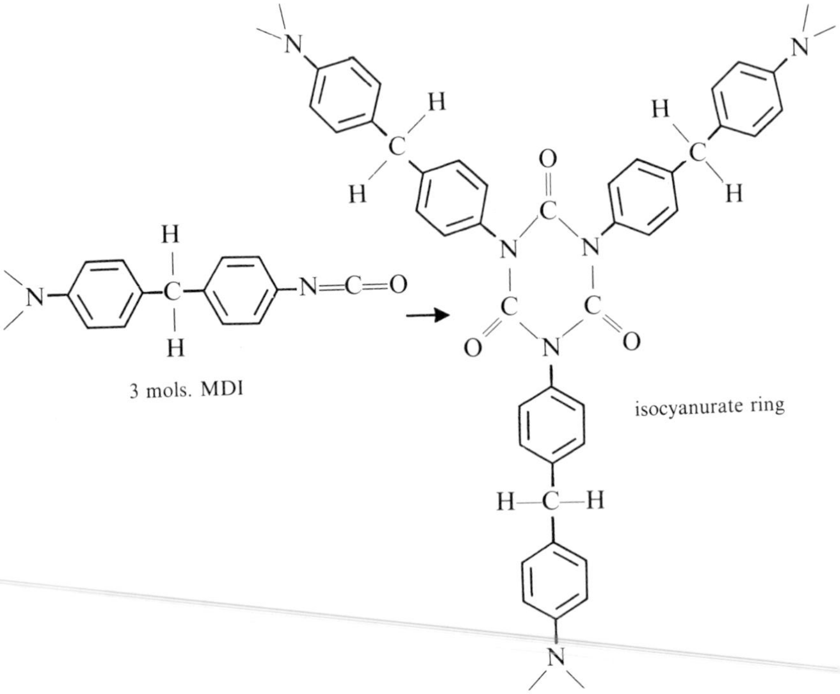

3 mols. MDI

isocyanurate ring

Isocyanurate foam from pure diphenyl methane diisocyanate is too friable to be of practical value and this necessitates the introduction of polyols bound in as urethanes. The initial urethane-forming reaction provides a useful source of heat to boost trimerisation and initiate vaporisation and foaming with trichlorofluoromethane. In practice some 10–30% of available isocyanate groups are converted to urethanes.

Hipchen[8] has published a bibliography of some of the significant publications on isocyanurate chemistry and has pointed out that the objective of formulating a urethane-modified polyisocyanurate is obviously to retain the greatest level of the desirable characteristics of each polymer, i.e. the superior thermal stability of isocyanurate and the lower friability and superior processability of urethanes.

Hipchen points out the need to have a system of expressing the level or concentration of isocyanurate, and suggests the following method:

$$\% \text{ isocyanurate} = \text{wt. }\%, \text{ free N} = C$$

in the polymer available for conversion to trimer

With crude MDI of equiv. wt. = 134 we have

$$\% \text{ isocyanurate} = \frac{42}{134} \times 100 = 31 \cdot 3$$

and with a formulation of 80·5 parts MDI of equiv. wt. = 134
6·1 parts polyol of equiv. wt. = 31

we have:

$$\% \text{ isocyanurate} = \frac{\dfrac{80 \cdot 5}{134} - \dfrac{6 \cdot 1}{31}}{80 \cdot 5 + 6 \cdot 1} \times 42 \times 100 = 19 \cdot 6$$

It is important to appreciate that just as the composition and performance of conventional rigid foams can be influenced by the type of isocyanate and polyol used so there can be substantial differences between isocyanurate foams depending not only upon the percentage of urethane groups present but upon the type of polyol and choice of trimerisation catalyst. Bechara[9] has reviewed the influence of several types of catalyst on the trimerisation and urethane reaction.

The superior thermal stability properties of isocyanurate foams depend upon their high aromatic content, the high degree of cross-linking and the high thermal stability of the isocyanurate ring system. Excessive 'dilution' with urethane groups must diminish these properties.

A thermogravimetric curve showing the substantial difference between 100% PIR foams, urethane group diluted PIR foams and TDI and MDI urethane foams was published by Ball et al.[3] and is reproduced as Fig. 1.

PIR foams have major advantages over conventional rigid urethane foams in respect of

1. Higher operating temperature, i.e. 140 °C compared with 100 °C;
2. Improved surface spread of flame resistance;
3. Reduced ignitability;
4. Less smoke development on burning;
5. Greater resistance to the US Bureau of Mines propane torch test; and
6. Improved fire resistance in composites compared to conventional urethane foams.

The improved performance of PIR foams in resistance to propane torch and fire resistance tests is because of the formation of a carbonaceous fibrillar network as a facsimile of the original foam structure. Once formed this char is destroyed only slowly in reducing atmospheres even at temperatures of the order of 1200 °C and it, therefore, acts as a flame and

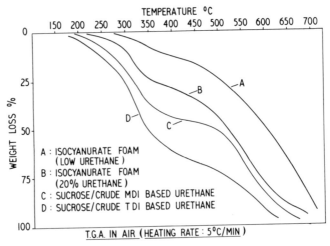

FIG. 1.   Thermogravimetric curves obtained in air.

heat barrier. Furthermore, the formation of a char means that less total heat is evolved compared with foams in which the total weight of foam contributes to the heat of combustion.

The superior performance of PIR foams compared with TDI and MDI based rigid urethane foams in small-scale tests such as ASTM D1692-68, the ASTM 3014/73 and the BS 476 Part 7 Surface Spread of Flame Test was reported by Ball et al.[4]

Ball et al.[3] reported experiments on 3 ft square panels subjected to the time–temperature curve laid down in BS 476 Part 8 (ISO R 834) for testing the fire resistance of structures. The results obtained are summarised in Table 1.

The data of Table 1 show that facings have a strong influence but comparisons of panels 4, 5, 7, 9, 10 and 11, respectively, indicate that the use of an isocyanate foam increases the failure time substantially.

Hipchen[8] has reported the commercialisation of a roof insulation product which employs isocyanurate foam reinforced with glass fibres and

TABLE 1

SIMULATED FIRE RESISTANCE TEST: BEHAVIOUR OF 3 ft SQUARE PANELS

Panel number	Exposed face	Unexposed face	Core Type	Core Thickness	Failure time[a] (min)
1	20 swg steel	20 swg steel	A	2	13
2	$\frac{1}{4}$ in asbestos insulation board	24 swg aluminium	A	1·75	13
3	$\frac{1}{8}$ in soft asbestos board	$\frac{1}{8}$ in soft asbestos board	A	1·75	15
4	$\frac{1}{4}$ in asbestos insulation board	$\frac{1}{2}$ in asbestos insulation board	C	1·5	19
5	$\frac{1}{4}$ in asbestos insulation board	$\frac{1}{4}$ in asbestos insulation board	A	1·5	20
6	$\frac{1}{4}$ in asbestos insulation board	$\frac{1}{4}$ in asbestos/cement board	A	1·5	25
7	$\frac{1}{4}$ in asbestos insulation board	$\frac{1}{4}$ in asbestos insulation board	B	1·0	29
8	$\frac{1}{2}$ in plasterboard	$\frac{1}{2}$ in plasterboard	A	1·0	39
9	$\frac{1}{2}$ in asbestos insulation board	$\frac{1}{2}$ in asbestos insulation board	A	1·0	46
10	$\frac{1}{4}$ in asbestos insulation board	$\frac{1}{4}$ in asbestos insulation board	B	2·0	52
11	$\frac{1}{2}$ in asbestos insulation board	$\frac{1}{2}$ in asbestos insulation board	B	1·0	59

[a] Reason for failure: Temperature Criterion
   A—conventional urethane foam.
   B—isocyanurate based foam.
   C—no core filling used.

surfaced with asphalt-saturated asbestos felt which is claimed to be the first totally cellular plastic product to pass the Factory Mutual Calorimeter and related tests required for FM approval as a component of Class I insulated steel deck construction. (Previously composites were necessary.)

Carbodiimide Foams

This type of foam has been described by Mann.[10]

Carbodiimide groups are formed by the use of special catalysts such as phospholanoxides partially substituted with aromatic/aliphatic side chains and/or halogens. The idea of rendering the catalyst complex with a ligand was an important breakthrough since it allows practicable mixing times. The reaction is

$$R-NCO + RNCO$$
$$\downarrow$$
$$R-N=C=N-R + CO_2$$

One mole of CO_2 is formed from every two NCO groups and foams the reaction mix. Using MDI, foam densities of the order of $16 \, kg/m^3$ are obtained. CO_2 has a very high rate of diffusion through cell walls compared to air and the thin walls collapse and give an open-cell foam. The reaction to form carbodiimide is only slightly exothermic and Mann reports that maximum temperatures of only $70 \, °C$ are observed in large blocks.

It is of interest to note that during the formation of carbodiimide bonds no increase in branching and cross-linking takes place in the polymer structure as it does in the trimerisation to isocyanurate foams. The stiffness of the carbodiimide foam is due to the cumulative double bonds.

Carbodiimide foams char when subjected to fire and generate less smoke than conventional urethane rigid foams. In the ASTM 3014/73 Test they give 'weight retained' figures similar to PIR foams, i.e. over 90% compared to PUR figures of approximately 30%.

Trichlorobutylene Oxide Based Polyols

Raymond[11] and Syrop[12] have reported substantially improved fire properties by the substitution of conventional rigid polyols with chlorine containing polyethers which utilise trichlorobutylene oxide. These polyols have the following structure and contain approximately 47% chlorine

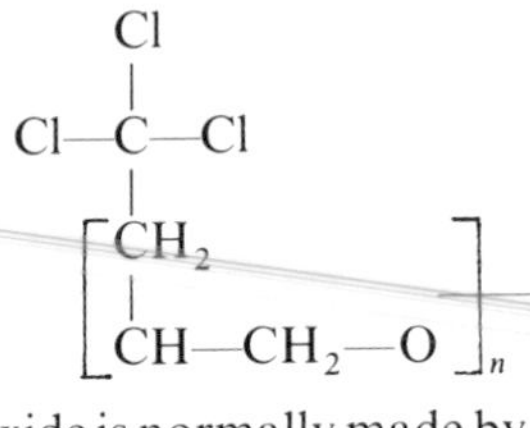

$$\begin{array}{c} Cl \\ | \\ Cl-C-Cl \\ | \\ \left[\begin{array}{c} CH_2 \\ | \\ CH-CH_2-O \end{array}\right]_n H \end{array}$$

Trichlorobutylene oxide is normally made by the reaction of allyl alcohol and CCl_4, followed by elimination of HCl

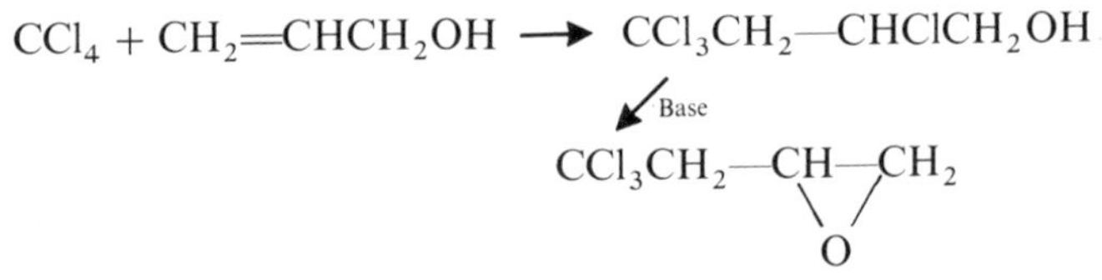

$$CCl_4 + CH_2{=}CHCH_2OH \longrightarrow CCl_3CH_2-CHClCH_2OH$$

$$\searrow^{Base}$$

$$CCl_3CH_2-CH\underset{\diagdown O \diagup}{-}CH_2$$

Allyl alcohol is, of course, expensive and this type of polyol is therefore more expensive than conventional rigid polyols.

Foams based on these polyols are superior to conventional PUR foams in resistance to surface spread of flame, smoke evolution and weight retained in the ASTM 3014/73.

Epichlorhydrin Based Foams

Solvay have developed[13] halogen-containing polyether polyols which give polyurethane foams with fire properties superior to conventional PUR foams. The polyether polyols have the structure

$$Z-[-\{-O-CH(CH_2Cl)-CH_2-\}_x-O-\{CH_2-CH(CH_2Cl)-O-\}_y$$
$$-CH_2-CHOH-CH_2OH]_z$$

The reactions leading to this type of structure are as follows:

$$R-OH + ClCH_2CHCH_2 \quad \text{(epichlorhydrin)}$$
$$\underset{O}{\diagdown\diagup}$$

$$\downarrow BF_3$$

$$R-O-\{-CH_2CH(CH_2Cl)O-\}_x-CH_2CH(CH_2Cl)OH$$

$$\downarrow \text{alkali}$$

$$R-O-[CH_2CH(CH_2Cl)O-]_x-CH_2CHCH_2$$
$$\underset{O}{\diagdown\diagup}$$

$$\downarrow \text{dilute acid}$$

$$R-O-[CH_2CH(CH_2Cl)O-]_x-CH_2CHOHCH_2OH$$

R = saturated or unsaturated polyol initiator

Polyimide Foams

Polyimide foams have been made by reaction of an acid anhydride with an isocyanate to give CO_2 which also blows the foam.

Alberino[14] has described the preparation and properties of polyimide foams in some detail. In particular he has pointed out they can be produced from isocyanates and anhydrides using a polyepoxide co-reactant which provides for room temperature or moderate temperature processing without the need for a solvent. The foams produced contain isocyanurate as well as imide groups and give a low smoke rating and excellent flame spread rating compared to PUR foams.

FUTURE POSSIBILITIES FOR CHEMICAL MODIFICATION

It is probably unrealistic to expect that massive improvements in the fire resistance of PUR or PIR foams are still possible by chemical modification of the polymer. The problem lies less in ignitability but more inherently in low density and high insulating power. Indeed, as the following figures show, ease of ignition as expressed by self-ignition temperature (ASTM D1929) is quite favourable compared with other materials commonly used in construction work.

	Self-ignition temperature ($^{\circ}C$)
PUR	approximately 500
Red deal	375
Bitumen	275

However, it is a fact that the burning rate of a low density material such as polyurethane foam is more sensitive to the rate of temperature rise at its surface than is the corresponding factor for a material such as wood. This property results in the foam burning weakly when exposed to a small flame and strongly when it is exposed to an intense radiant source. The phenomenon can easily be seen by the differences in the weight loss of PUR foam and wood when exposed to say a 1 kW and 5 kW heat flux, reported by Roberts.[15]

Roberts points out that the rapidity of ignition of the foam is caused by its thermal properties. The rate of surface temperature rise of a material supplied with heat at a constant rate from an external source is inversely proportional to the product (Kpc) where K = thermal conductivity, p = density and c = specific heat. Since thermal conductivity is approximately proportional to density, (Kpc) is extremely sensitive to the density of a material.

The ratio of (Kpc) for a low density polyurethane foam ($p = 40\,\text{kg/m}^3$) to (Kpc) for a typical hardwood ($p = 600\,\text{kg/m}^3$) is approximately $1:100$ so that a polyurethane foam surface will be heated locally to its ignition temperature by a flame in 1/100 of the time required to heat a wood surface in the same way. Also, the spread of flame across a surface may be regarded as a progressive ignition process since new areas of surface are continuously being raised to the ignition temperature.

In his paper, Roberts has pointed out that several of the characteristics of burning urethane foam are a natural consequence of the physical structure of the substance of the foam typified by the estimated value of $E = 270 \, \text{kJ/mole}$ for the activation energy of the pyrolysis reactions. Many polymeric materials have a similar value for E and might be expected to display similar characteristics. The lower the density and the thermal conductivity of a flammable material the more rapidly can it be ignited and the higher will be the rate of flame spread across an exposed surface of the material.

It is important to realise, however, that these comments on the theory of the behaviour of foam to radiation sources refer to uncovered foam. They are relevant, therefore, to a surface of sprayed foam but are not relevant to foam covered with a facing material. The great majority of the applications of rigid foam are as a composite and it is the fire resistance properties of the composite which should therefore be our main concern.

RIGID FOAM IN COMPOSITES

In practice, rigid foams are rarely used alone but rather in the form of composites. In most building applications the foams are covered by or directly laminated with plasterboard, steel, bitumen felt, etc. The ability of rigid foams to adhere to a wide variety of facing materials to give an integral high strength/low weight structure is of importance in building design terms. Some critics appear to be unwilling to judge urethane foams on the basis of their fire performance as used but would, for instance, dismiss as nonsense a judgement of the fire performance of a 3 in plank of wood based on tests of the flammability of wood shavings.

Composites affect not only the ease of ignition of the foam but the speed of development of the fire and the rate of release of toxic gases. It is sometimes possible to have a situation where a foam which appears inferior when tested in the uncovered form may perform better in a composite.

In the work carried out by ICI, approximately 8 years ago, prototype buildings constructed of urethane rigid foam laminate were tested in a large-scale fire. It is not proposed to discuss the details of this work which

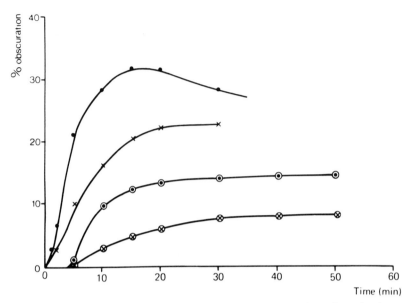

FIG. 2.   Increase in light obscuration by smoke generated from foam laminates DD36:1974 method. ● paper faced polyurethane foam; ◉ paper faced polyurethane foam + 3 mm plaster; × paper faced polyisocyanurate foam; ⊗ paper faced polyisocyanurate foam + 3 mm plaster.

are already published[4] but merely to draw out from these tests the difference between the performance of a composite and foam itself. In the test, two test buildings were constructed from urethane foam sandwiched between metal facings and a third building used as control building was lined with traditional materials. All the buildings, instrumented for both temperature and gas analysis, were subjected to a substantial wood crib fire. One of the results of great significance was the appearance of the steel/foam composite when opened up after the fire. In spite of the intensity of the fire, most of the foam remained undamaged.

The generation of smoke and combustion gases from foam are also substantially affected by composite design. Work has been described[16] in which samples of uncovered foam and foam laminated with plasterboard were burnt and the products of combustion allowed to fill a large room as described in the British Standard 'Draft for Development' No. 36 (now withdrawn). Figures 2 and 3 show the results. Figure 2 shows that if, instead of using polyurethane, polyisocyanurate is used, there is an improvement evident in terms of smoke emission even with the bare foam itself. When the foams are put behind plasterboard in turn the levels of obscuration are substantially reduced under the same conditions. In this work the atmosphere was sampled and measurements made of carbon monoxide and hydrogen cyanide concentrations. Figure 3 shows the concentration of carbon monoxide as a function of time. The least favourable case is again paper faced urethane foam, lower down there is polyisocyanurate foam and

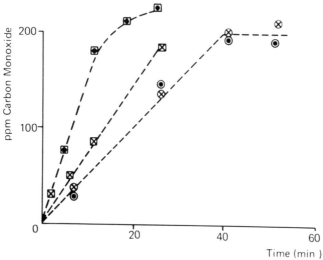

FIG. 3.   Levels of carbon monoxide in test room during DD36:1974 smoke test. ▣ paper faced polyurethane foam; ◉ paper faced polyurethane foam + 3 mm plaster; ⊠ paper faced polyisocyanurate foam; ⊗ paper faced polyisocyanurate foam + 3 mm plaster.

when both these types are placed behind plaster the performance becomes very similar and the carbon monoxide levels are at their lowest. A similar order of reduction was obtained when comparing hydrogen cyanide evolution.

Zorgmann[17] has reported data which support the argument that tests on rigid foams themselves are not necessarily relevant to their performance in composites. In a paper entitled 'Behaviour of Insulated Steel Roofs in Fully Developed Fires' he reported on some 1:1 *ad hoc* tests carried out at the request of the International Isocyanates Institute.

Tests were carried out on steel roofs insulated with various insulating slabs and covered with a roof covering material. The roofs were mounted on a test building constructed from gas concrete slabs. The building consisted of two symmetrical rooms measuring 3 m × 3 m and height 2·46 m separated by a dividing wall. In one of the rooms a fire load of 30 kg/m² of wood cribs was burned.

Zorgmann points out that the moment at which the integrity-limits (involving the outside of the roof) are reached in a fully developed fire can be expected to depend upon the construction of the insulated steel roof, the insulating value at elevated temperatures of the insulating slabs and the tendency to develop combustible gases at temperatures reached during the fire by insulating—and roof covering—materials.

In the 1:1 test each roof was covered with two layers of bitumen bonded glass-fibre sheet. The first layer, comprising perforated sheet, was loosely laid on top of the insulating slabs. The mineralised top layer was glued to the first layer and through the perforations to the insulating slabs using 30 kg bitumen for each roof.

To be able to compare data on ignition of and fire spread over the roofs for differently developing fires, times were corrected for the mean flash-over time of all tests. Zorgmann reports that the ignition times of roof coverings depended upon not only on the insulation value at elevated temperature of the foam but also on the tendency to deformation, and therefore thermoplastic foams are at a disadvantage to thermosetting foams in this respect.

Zorgmann's figures for fire spread show very little difference in performance between polyisocyanurate foams, P/F foams and mineral wool; all gave longer times than the polystyrene foam tested.

Zorgmann also compared the results obtained in the 1:1 roof tests with results obtained on the insulating materials with the standard Dutch fire test for surface spread of flame and flash over. His conclusion was as follows:

> No correlation was found between the fire behaviour of the roof as a whole and the classification of bare surfaces of the insulating slabs themselves in standard fire tests approved to predict the performance of wall and ceiling linings in developing fires. An interesting conclusion because especially fire insurance companies in several countries base their judgement of the fire risk on such classifications.

The SPI Urethane Safety Group in their Bulletin U102, dated November 1974, have also noted that, for polystyrene foams, large-scale combustibility behaviour in the Corner Wall Test was not predicted by the numerical ratings as determined on foam by ASTM E-84. This Bulletin reports the results obtained in Corner Wall Tests by a number of composites.

More recently Nadeau[18] and Nadeau et al.[19] have reported on a comparison and evaluation of results obtained on exposed and coated rigid urethane foam when subjected to large-scale corner and compartment testing. These are two important papers since they discuss (a) the ability of the FM Large-Scale Corner Wall Test to predict fire behaviour not only in large room situations such as in industrial buildings but also of similar materials in the wall/ceiling applications of residential and municipal buildings, and (b) the correlation between the Large Scale Corner Test and the Underwriters' Laboratory (UL) Room Compartment Corner Test.

Nadeau concludes that combustible rigid spray on foam can be effectively protected with cementitious coatings of the type investigated and reports good correlation for finish rating value in both Corner Wall Test and the UL test.

SMOKE EVOLUTION

Reference has been made earlier to the effect upon the amount of smoke evolution by composite design. There is an assumption by some people that visible black smoke is the criterion by which the ability to see and escape from a building is judged. This view can be quite misleading since some smoke, even if it does not appear to be black and does not give high smoke readings on instruments, can in fact contain lachrymators which will render a human being temporarily blind. Such typical lachrymators are acrolein and formaldehyde and before reaching conclusions on the difference in smoking properties of materials in a fire it is essential that any potential lachrymatory effect is also taken into account.

Comparisons of various types of foam can also be misleading if their method of use is not also considered. In the work by Zorgmann[17] on roof structures, in which profiled steel laminate containing various types of foam and mineral wool were subjected to a substantial fire within the building structure, the effect of the foam on smoke evolution is insignificant compared with the performance of the total composite. In this type of roof structure there was a covering of bitumen and the contribution of the insulating materials to smoke evolution was insignificant compared to that evolved when the bitumen flashed. Figures 4 and 5 give some indication of the difference between the 'before' and 'after' situation.

In many types of roof there is a first layer of $4 \, kg/m^2$ of bitumen, then a layer of $1 \, kg/m^2$ of urethane and then $8 \, kg/m^2$ of bitumen felt. The

FIG. 4.   Smoke evolution—before bitumen flash.

FIG. 5.   Smoke evolution—after bitumen flash.

comparison in calorific value and smoke evolution is therefore of 12 kg/m²
of bitumen against 1 kg/m² of urethane in this type of fire test.

Christianson and Waterman[20] have reported that several types of room
and corridor tests correlated poorly to the XP-2 chamber and NBS
chamber and the 25 ft tunnel, the latter correlating best. These authors
commented that the average of the flaming and non-flaming mode in the
NBS chamber correlated better than either mode alone. They also
suggested that smoke ratings should be combined judiciously with flame
spread rates to obtain a truer indication of full-scale behaviour. The writer
believes this suggestion is also valid when comparing toxicity ratings.

TOXICITY OF SMOKE

It is all too readily assumed that the presence of a toxic substance in gases
evolved in a fire thereby constitutes a hazard. With the development of
highly sensitive techniques of chemical analysis a very great number of
different products may be identified in fire gases, but what really matters is
'how much' and at what rate it is produced. 'Toxicity', which is an absolute
property of a chemical, must not be confused with the 'toxic hazard' it *may*
present in the fire situation. It should be remembered that carcinogens are
present in domestic bonfires and grilled steak and that when tobacco
undergoes combustion very many different toxic compounds are present in
the smoke including CO, HCN, acrolein, aldehydes and nicotine.

Even using small-scale tests with animals to assess the presence of toxic
substances can lead to misleading conclusions.

Cornish *et al.*[21] have, for instance, warned that the premature adoption
of methodology for toxicity testing may do more harm than good, pointing
out that results can be altered by conditions of pyrolysis, oxygen supply,
heating conditions and by the end point chosen (e.g. mortality, lung
pathology, behavioural effects). Cornish compared the LC_{50} value for a
variety of polymers under static chamber (rapid combustion) and dynamic
chamber (slow pyrolysis) conditions. Table 2 shows one set of such
comparisons. It is of interest not only in showing the effect of combustion
conditions but also in indicating that on this evidence polyurethanes are no
more toxic than many other materials.

There are a number of approaches to the generation of toxic gases by
burning materials for animal testing. Some of these have been described in
the literature and it is not the intention in this chapter to consider these in
any detail. West Germany is the only country where the same equipment

TABLE 2
COMPARATIVE MORTALITY DATA OF COMBUSTION PRODUCTS OF POLYMERS

Static chamber		In order of decreasing toxicity	Dynamic chamber	
LC_{50} (g)	Sample		Sample	LC_{50} (g)
9	Red Oak (most toxic)	1	Wool	0·4
10	Cotton	2	Polypropylene	0·9
21	ABS (FR)	3	Polypropylene (FR)	1·2
23	SAN	4	Urethane foam (FR)	1·3
25	Polypropylene (FR)	5	Polyvinyl chloride	1·4
28	Polypropylene	6	Urethane foam	1·7
31	Polystyrene	7	SAN	2·0
33	ABS	8	ABS	2·2
37	Nylon 66	9	ABS (FR)	2·3
37	Nylon 66 (FR)	10	Nylon 66	2·7
47	Urethane foam (FR)	11	Cotton	2·7
50	Urethane foam	12	Nylon 66 (FR)	3·2
50	Polyvinyl chloride	13	Red Oak	3·6
60	Wool (least toxic)	14	Polystyrene	6·0

(draft DIN 53436) has been used by the majority of researchers in this complex field.

Figure 6 shows the DIN 53436 equipment in diagrammatic form. It shows both the combustion chamber and the exposure chamber where the animals breathe an atmosphere containing the products of combustion. Tests are carried out under standard conditions, with several parameters capable of variation. The most important of these is the temperature of combustion, which is effected by an annular electric oven, which moves at a predetermined rate over the sample contained in the combustion chamber. A number of different temperatures are used, at steps of 50 °C. The range of pyrolysis temperatures between that level which reproducibly causes no deaths and that level which reproducibly does cause some deaths of the exposed animals can then be determined. This temperature is sometimes called the 'critical temperature'.

Using the DIN 53436 apparatus Kimmerle[22,23] and others have obtained comparative critical temperatures for a number of materials. Some of the results are summarised in Tables 3 and 4. From this work Kimmerle has concluded that, in terms of toxicity of pyrolysis products, PUR foams compare favourably with many conventional materials they replace.

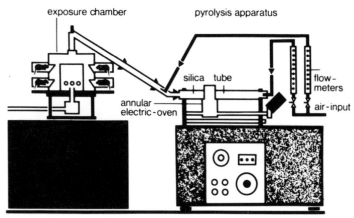

FIG. 6. Arrangement of apparatus for DIN 53436.

In recent years there has been much debate on the various methods available or being developed to use animals to assess toxic hazard from smoke. The Flammability Research Center at the University of Utah has published a great deal of experimental data on this topic[24,25] and has argued that the objective should be to develop more sophisticated techniques for assessing hazard.

Studies at Utah embrace the gathering of fire statistics, development of analytical methods for pyrolysis products and an assessment of the

TABLE 3

COMPARISON OF THE RELATIVE TOXICITIES OF THE PYROLYSIS PRODUCTS OF SYNTHETIC AND NATURAL BUILDING MATERIALS TESTS WITH EQUAL VOLUME

Materials	Lowest temperature ($^\circ C$) indicating mortality
Rigid urethane foam	600 and >600
Rigid isocyanurate foams	500, 600 and >600
Semi-rigid polyurethane foams	500, 600 and >600
Foams based on UP resins combined with expanded glass beads	>600
ABS	400
Polycarbonate	600
Spruce-wood	350
Cork	300

TABLE 4
COMPARISON OF THE RELATIVE TOXICITIES OF THE PYROLYSIS PRODUCTS OF SYNTHETIC AND NATURAL MATERIALS USED BY THE FURNITURE INDUSTRY TESTS WITH EQUAL WEIGHTS

Materials	Lowest temperatures (°C) indicating mortality
Flexible urethane foam (Polyether)	300 and 400
Flexible urethane foam (Polyester)	>450
High resilience foam	500
Coconut fibre	250
Rubberised hair	300
Latex foam	>200
Flocks	300
Lint wool	300
Wool	300
Cotton	250

physiological effects of gases formed in fire situations. This has included studies of behavioural response which some experts argue is a most important factor in assessing the hazard of gases or combinations of gases.

Also of great interest are the results of the University of Utah study on the effects on test animals of pyrolysing Douglas Fir. They have shown that under conditions of low heat flux the wood elicits a strong toxic influence on laboratory test animals leading to rapid incapacitation and death. Examination of the animals after exposure showed severe attack upon the lung tissue and thyroid gland and death was not attributable to carbon monoxide evolution but to aldehydes, ketones, acroleins, etc. This type of data on natural materials shows that hazards from synthetic materials are not unique and allows a more meaningful assessment of what additional hazards, if any, are represented by rigid urethane foams.

Further information comparing the toxicity of synthetic and natural materials has been published by Hilado and co-workers.[26,27]

As stated earlier, it is essential to keep in mind that in a real fire situation ignitability and fire spread and therefore rate and concentration of gas evolution are the principal parameters determining hazard. However, the small-scale animal experiments may allow the presence of unknown 'super toxicants' to be detected and should enable an assessment to be made of the effect of formulation changes.

POSSIBLE FURTHER ACTION BY THE RIGID FOAM INDUSTRY

Some of the work carried out by the industry to improve the fire performance of rigid foams has been described above. There is, however, no point in expanding effort on improved products and on obtaining data which helps define optimum design for fire safety unless this data is communicated effectively. The total industry both in the UK and overseas could probably improve their performance in this respect.

This activity should not be confined to a negative defensive attitude on the fire properties of rigid foams. There is a need to draw the attention of government, the public and the media to the benefits available from the superb insulation properties of urethane when used in refrigerators, building, refrigerated transport, etc.

It is also important to bring home to government and the public the benefits that rigid urethane foam insulation offers in terms of energy conservation. Dr Martin Bloom has argued[28] the need for long-term solutions to the energy crisis by substantially raising the standards of insulation in the northern hemisphere. He has pointed out that an extra £300 spent on improving the standard of insulation in the average home would halve the amount of energy used in domestic heating in the UK. He estimated the cost of heating homes, offices, factories, schools, etc. throughout Britain would be expected to rise from the current figure of £3000 m a year to something like £10 000 m by 1984. Dr Bloom suggests that improved insulation would not only give substantial savings of capital and operating costs but also the environment would be spared a substantial number of nuclear power stations which will otherwise be erected to provide the energy required.

Other insulation materials could of course also be used to achieve these savings; there are good reasons for arguing that urethane rigid foams have a significant part to play.

REFERENCES

1. EINHORN, I. N. and NEWMAN, M. (1975). Fire Injuries: Case History Studies, Flammability Research Center, University of Utah.
2. STARK, A. W. (1976). *Chemistry and Industry*, 17th Apr., 359.
3. BALL, G. W., HAGGIS, G. A., HURD, R. and WOOD, J. F. (1968). *J. Cell. Plast.*, **4**(7), 248.

198   R. HURD

4. BALL, G. W., BALL, L. S., WALKER, M. G. and WILSON, W. J. (1971). *J. Cell. Plast.*, **7**(5), 241.
5. BIRKY, M., EINHORN, I. N. *et al.* (1973). Flammability Research Center, University of Utah, FRC/UU-14-UTECH 73-178.
6. EINHORN, I. N. *et al.* (1976). *Proc. SPI 4th International Cellular Plastics Conference*, Montreal, 15–19th Nov., 217–57.
7. WEIL, E. D. and AARONSON, M. A. (1976). Conference—Recent Advances in Combustion and Smoke: Retardance of Polymers, University of Detroit, 25–27th May.
8. HIPCHEN, D. E. (1976). *Proc. of SPI 4th International Cellular Plastics Conference*, Montreal, 15–19th Nov., 217.
9. BECHARA, I. (1976). *Proc. of SPI 4th International Cellular Plastics Conference*, Montreal, 15–19th Nov., 207.
10. MANN, F. (1975). 5th International Conference on Plastic Foams, Dusseldorf, 25–26th May.
11. RAYMOND, M. A. (1976). *Proc. of SPI 4th International Cellular Plastics Conference*, Montreal, 15–19th Nov.
12. SYROP, H. *Advances in Urethane Science and Technology*, Vol. 4, Technomic Publishing Co., USA.
13. UK Patent 1,371,489.
14. ALBERINO, L. M. (1976). *Proc. of SPI 4th International Cellular Plastics Conference*, Montreal, 15–19th Nov., 1.
15. ROBERTS, A. F. (1974). *Insulation*, **18**(6), Nov./Dec., 10.
16. BALL, G. W., BALL, L. S., WALKER, M. G. and WILSON, W. J. (1972). *Plastics & Polymers* (Reprint), **40**(149), Oct., 290–303.
17. ZORGMANN, H. (1976). V.F.D.B. 5th International Fire Protection Seminar, Karlsruhe, 22–24th Sept.
18. NADEAU, H. (1976). *Proc. of SPI 4th International Cellular Plastics Conference*, Montreal, 15–19th Nov., 198.
19. NADEAU, H., DARR, W. and HOFRICHTER, C. (1977). *J. Cell. Plast.*, **13**(2), 102.
20. CHRISTIANSON, W. and WATERMAN, T. (1971). *Fire Technology*, **7**(14), 332.
21. CORNISH, H. H. *et al.* (1976). International Symposium on the Toxicity and Physiology of Combustion Products, University of Utah, Mar.
22. KIMMERLE, G. (1974). *J.F.F. Combustion Toxicology*, **1**.
23. KIMMERLE, G. (1974). Physiological and Toxicological Aspects of Combustion Products, International Symposium, University of Utah, Mar.
24. PETAJAN, J. (1976). International Symposium on the Toxicity and Physiology of Combustion Products, University of Utah, Mar.
25. PACKHAM, S. (1976). International Symposium on the Toxicity and Physiology of Combustion Products, University of Utah, Mar.
26. HELADO, C. J. *et al.* (1976). *J. Combust. Toxic.*, **3**(4), Nov., 345–62.
27. HELADO, C. J. *et al.* (1977). *J. Combust. Toxic.*, **4**(1), Feb., 16–68.
28. BLOOM, M. (1974). *ICI Magazine*, **52**, Nov., 242–6.

Chapter 8

NEW APPLICATIONS FOR POLYURETHANES

R. HURD

ICI Organics Division, Manchester, UK

SUMMARY

Selected recent developments in urethane technology are reviewed. The subjects chosen are (a) the use of isocyanates in particle board manufacture, (b) one-component froth foams, (c) structural rigid foams and (d) urethane carpet backing.

INTRODUCTION

To survey all the new developments in urethane technology in one chapter would be an impossible task. Some developments such as improved equipment for flexible slabstock and new lamination techniques for rigid foam are described in Chapters 4 and 6 of this book. In this chapter the four developments described have been chosen because, hopefully, they illustrate the versatility of the chemistry and processing characteristics of urethanes.

Thus the use in particle board illustrates the use of isocyanates as a minor constituent substantially modifying the properties of an established material, whilst use in one-component froth illustrates an application where the isocyanate is a major constituent and where the objective was to develop a method of introducing an insulating gap sealant for energy conservation without the need to operate metering and mixing equipment.

The sections on structural rigid foams and carpet backing illustrate how urethane chemistry and processability have achieved, in one step, technical effects previously requiring two or more stages. Thus in structural foams a

199

sandwich structure can be obtained in one operation and, in carpet backing, urethane flexible foams can provide in one step the adhesive, tuft lock and foam back.

URETHANE BINDERS FOR PARTICLE BOARD MANUFACTURE

During recent years there has been an increase of interest in the use of isocyanates as binders for wood-chip board and particle boards based on flax, straw, etc.

At the early stage of their industrial development isocyanates were used as adhesives and wood and plywood were among the materials on which work was done.[1] Isocyanate adhesives for wood were of interest because the hydroxyl groups in the cellulosic and phenolic components in wood are available for chemical cross-linking and bonding.

The binders used in the commercial development of wood-chip board were, of course, urea formaldehyde (U/F) resins which were available as easily applied aqueous syrups and were cheap.

World production of particle board, which started in the mid-1942s, has grown at a rate of 15 to 20% to reach the following levels by 1972:[2]

	Million m^3		
	1970	1971	1972
Plywood + blockboard	32·7	36·1	39·0
Particle board	18·3	22·0	25·9
Fibre board	7·9	8·4	9·1

As the usage of U/F bonded wood particle boards grew in furniture and building applications, it became clear that they were not suitable for some building applications due to susceptibility to attack by moisture.

Phenol formaldehyde (P/F) bonded boards have been developed which show significantly better retention of mechanical properties under wet conditions. In Europe the density of P/F boards is normally 5 to 10% higher than the U/F bonded boards in order to meet official building requirements. P/F boards are, therefore, substantially more expensive than the U/F bonded boards. They have, however, achieved widespread use in the building industry and production is about 10–15% of the total.

Deppe[3] has pointed out that the use of MDI as an adhesive in particle board manufacture offers certain advantages over P/F bonded boards. The density of MDI boards may be lowered by 10–15% without an unacceptable reduction in their strength properties and press times may be reduced as no additional water need be introduced into the mat. Deppe points out that under present manufacturing conditions the moisture content can be so critical that an excess of only 0.5% can cause the production of cracks due to the excessive steam pressure in the board during pressing. The use of isocyanates may offer a solution to these problems.

Deppe[3] reports that isocyanate bonded particle boards give equivalent weather resistance to phenolic boards, and superior strength perpendicular to the surface under wet and dry conditions. Some test results indicate that MDI can be used to produce boards of very satisfactory properties when used at 5–6% on the weight of particles compared to an average of 10% for conventional binders. This closes the price gap between P/F and MDI binders.

The use of isocyanate binders may expand raw material feedstocks considerably. The processing of flax and straw with P/F or U/F binders has not yet achieved boards with satisfactory moisture resistance and it would appear probable that improved properties are achieved when MDI is used to glue these products.

Disadvantages of Urethane Binders

The principal problem in using isocyanate binders has been the sticking of the facing layers of the particles to the metal cauls, platens, etc. of the particle board presses.

Conventional release agents such as paraffin compounds were not satisfactory and silicones presented an additional problem for surfaces which were subsequently to be painted. Cauls and platens treated with a material such as Teflon or plastic films were not satisfactory because of high initial cost and damage by particles.

Until now the limited production with urethane binders has been carried out either by making veneered board, the facing being bonded during pressing thus obviating the need for release agents, or by using isocyanate binders for the core layer only and a conventional P/F binder for the facing layers. This has been acceptable since it is normal practice for the face layers to be made from smaller particles with a higher binder content than the core material. Hence the particle board plants are normally equipped with two streams of particles having different binder contents and it is only an extension of this practice to use different binders.

Deppe has reported that new release agents are being developed and states that a stearate release agent developed by ICI appears to be very promising. This release agent may be applied in water sprays onto the lower caul or steel belt and onto the top of the formed mat. The release agent transfers to the top platen during pressing. Deppe reported satisfactory performance of this release agent at an application rate of $40 \, \mathrm{g \, m^{-2}}$.

Progress has also been made in the application of isocyanates by the use of aqueous MDI emulsions.[4] These permit existing gluing equipment to be used and the low viscosity of the emulsion gives a better distribution over the particles.

From the above summary it will be seen that although present consumption of MDI as a particle board binder is small many of the technical problems involved in its use have been or are being resolved, and that property advantages for some applications should lead to a substantial usage of isocyanate binders for chipboard in the future.

ONE COMPONENT FROTH

One component froth (OCF) is a self-expanding, self-adhesive moisture curing gap filler, developed by ICI.[5,6] Chemically it is an isocyanate ended prepolymer blended with surfactant and catalyst and mixed with di-fluoro-dichloromethane which acts both as a frothing agent and a propellant.

The OCF is dispensed from a simple pressurised container or aerosol-type can and requires no mechanical or power source. When the valve of the container is opened the contents of the cylinder emerge from the nozzle as a well-expanded froth which is easily directed into a gap or hole. Specially designed valves allow dispensing to be well controlled with a simple on/off action.

The foam, as it leaves the cylinder, has the consistency of a shaving cream, is very sticky and will adhere firmly to most surfaces without any pretreatment. Adhesion to damp surfaces is excellent, a particularly important point on a building site. Once in the gap the foam expands to about double its volume.

During curing of the OCF the isocyanate ended prepolymer reacts with moisture in the air or from the substrate, producing substituted urea groups and carbon dioxide, the latter giving the foam expansion. The final foam volume is dependent upon the formulation of the prepolymer (particularly catalyst content), temperature and other factors but is normally up to twice the volume of the foam extruded from the nozzle pressure container.

When it is cured the foam may be trimmed with a sharp knife and finished by normal plastering, papering or grouting techniques. It can also be recessed on external surfaces to accept a conventional sealant to provide weather resistance.

The flexibility of the foam can be modified by the choice of polyol. As with all other prepolymer manufacture it is important to keep the polyol dry and to ensure that at all stages of preparation there is an excess of isocyanate present.

In many countries the cylinders used to contain the finished product are subject to statutory regulations. There are a number of specifications relating to the design of pressure vessels. Amongst the most widely recognised are BS 401 and ICC 4B 300 which cover 'low pressure gas vessels'.

It is important to avoid extremes of temperature when storing OCF cylinders since there is a progressive separation of di-fluoro-dichloromethane into a separate phase at the bottom of the vessel due to distillation effects. Even slight separation can show as pockets of gas emerging from the nozzle which break up the smooth extrusion of the froth. These phases are not easily re-mixed; hence, it may require a fair amount of agitation to achieve homogeneity.

OCF can be dispensed at temperatures down to sub-zero but under these conditions the pressure container and its contents must be at a temperature of at least 20 °C. Best results are obtained, however, when the materials are dispensed at temperatures around 20 °C to 35 °C and relative humidities above 40 %. The % humidity influences the rate of cure.

Physical Properties

The physical properties obtained will obviously vary depending upon the prepolymer formulation chosen. It is not easy to obtain physical properties of OCF foams by conventional methods since attempts to extrude test samples of large cross-sections lead to coarse foam and semi-collapse in the centre of the foam due to the long period of time before moisture can diffuse to the core and cure the foam. The general level of results obtained are as follows:

Dispensed froth density	$= 60\text{--}70 \, \text{kg/m}^3$
Cured foam	$= 30\text{--}45 \, \text{kg/m}^3$
Initial closed cells	$= 60\text{--}90 \, \%$
Water vapour transmission	$= 3\text{--}6 \, \text{perm/ins.}$

Applications

In Europe and the USA there is an increasing use of OCF because of its simplicity in use and versatility in application. The ability to apply the froth without metering equipment and in a wide variety of weather conditions is an important advantage. The high viscosity of the froth emerging from the container also allows application to vertical and overhead surfaces. This property is particularly valuable when using OCF as a gap sealant or adhesive in construction work (e.g. bonding sheets of plywood to masonry or concrete walls).

The 'energy crisis' has led to an expanding use of OCF as a gap filler/insulating material. The effect of air infiltration upon the energy consumption in residential construction is now becoming widely recognised[7] and there is increasing interest in this simple method of sealing off openings and gaps around windows, door frames, etc. The ease of applying OCF to gaps in vertical surfaces and its ability to expand and cure to form an insulating sealant is leading to a substantial growth of interest in this particular method of using urethane technology.

STRUCTURAL RIGID FOAMS

Simulated Wood

At the early stage of the development of rigid urethane foams in the early 1960s, high density foams were produced for specific technical outlets such as specialised applications in aircraft, load-bearing insulated floors, etc. In addition, a substantial outlet developed, particularly in North America and Italy, as a wood substitute.[8]

Compared to molten thermoplastics the unreacted urethane mix has a low viscosity and allows more easily the accurate reproduction of fine surface detail. Mouldings up to approximately $500 \, kg/m^3$ could be produced which had the appearance, feel and sound of wood! These properties, plus the relatively low capital investment required for the production of limited numbers of large mouldings, led to the use of the high density foams in simulated wood cabinets and for the decorative fronts of the so-called Spanish styled furniture. In some designs the decorative fronts were mounted on plain timber doors, etc., to form a composite.

The equipment used for foam production was the conventional high or low pressure urethane dispensing machines. Moulds were usually made from silicone rubber or urethane elastomer in order to facilitate the introduction of undercuts and to obtain accurate reproduction of fine surface detail. The most popular finishing technique was to coat the interior of the mould with a barrier/release coat pigmented to the base colour of the wood being simulated.

Properties are dependent upon the density and formulation chosen but in general are adequate for the end uses for which they are employed. In particular, indentation hardness and scratch resistance are equivalent to most natural woods. These properties are significant in decorative panels where resistance to scratches and small dents is advantageous. A compromise has normally to be made in balancing other properties, formulations giving higher heat distortion temperature having lower impact strength.

Wood has one major property advantage—its behaviour in flexure, giving higher breaking stresses and moduli.

Integral Skin Rigid Foams

The range of possible uses and applications of high density rigid foams was greatly enlarged by the development in Germany by Bayer of a process for the moulding of high density, integral skin structural rigid foams which were named Duromers. This process produces an integral skin moulding in a single operation providing mouldings with an unexpanded hard durable smooth outer skin with a lower density core.

In contrast to the earlier simulated wood technique, the mould cavity is charged with 50 to 150 % more material than would normally fill it when expanded under free-rise conditions. The internal pressures generated on the face of the tool (ca 3 to 6 kg/cm^2) cause liquefaction of the trichlorofluoromethane, thus producing a dense outer skin.

In the early stages, expansion of the liquid in contact with the mould surface is inhibited by the latter's thermal conductivity and thermal capacity. Expansion occurs in the core as the exothermic heat develops during the formation of the urethane polymer and the tool is filled.

Polymer expansion is usually achieved by a combination of carbon dioxide and fluorochloro hydrocarbons. The CO_2 is formed by the introduction of a small amount of water in the formulation and mixed blowing is usually preferred since water helps to achieve good flow properties and trichlorofluoromethane is beneficial in producing thicker skins. It is now commonly accepted that the description 'polyurethane structural foam' is suitable for the integral skin products of this type.

When first introduced the structural rigid foams were considered mainly for short production runs or for large mouldings because of the long cycle times, raw material wasted as sprue and reject rate. Of particular interest at this time was the work carried out by British Leyland on experimental structural foam bodywork. Some of the problems to be overcome in the use of plastics in automobile bodies have been discussed by Hill[9] who listed some of the technical requirements and compared the tool costs, material costs and labour costs of a steel body against those for thermoformed sheet, rotationally moulded thermoplastics and urethane structural foams.

Subsequently, Hill[10] reported the moulding of an experimental BLMC Mini-car in structural foam. Satisfactory performance in road tests was obtained and composite mouldings incorporating metal structures designed to provide stiffness, and protection against roll-over were developed.

Since 1972 advances have, however, been made, both in improved versions of the urethane systems offered by suppliers and in the availability of automated high-speed processing equipment.[11,12]

Development work on equipment was accelerated by the growing interest of major auto companies in the reaction injection moulding (RIM) process for micro-cellular auto bumpers, etc. described in Chapters 4 and 5. The introduction of multi-mould urethane lines fed from one metering unit, the use of multi-cavity tooling and the development of urethane systems containing internal release agents have made structural rigid foams competitive with thermoplastic injection moulding processes for some applications, e.g. components weighing more than 2 to 3 kg or runs producing less than 50 000 pieces.

Modern production lines offer a process which can operate, sequentially, a number of moulds of different sizes and complexity within a shot weight of 0·5 to 25·0 kg and sections of differing thickness can be produced in the same mould without sink marks.

Jacobs[12] has reported the investment required in thermoplastic structural foam processes and RIM type urethane processes and compared the production costs in structural urethane and polypropylene for a run of 240 000 pieces per annum of a structural plastic part with a volume of 2·6 litres. This calculation shows a cost advantage for the urethane system. Jacobs reports that an even greater cost advantage is shown for urethanes if the structural plastic has to be formulated to meet the fire retardancy specification of the Underwriters' Laboratory 94V-O.

Chemical system development has aimed at reducing demould time, raising heat distortion temperature and ensuring greater reproducibility in production. Reduced demould times and improved heat distortion temperature (HDT) have been achieved by polyol formulation (e.g. the substitution of sucrose or aromatic ring based polyol for trimethylpropane based polyols), and/or by the introduction of some isocyanurate groups into the polymer. During recent years demould times have been reduced from 5–10 min to 2–4 min and HDT raised from ca 80 °C to 180 °C (DIN 53424).[13]

One of the major causes of the early difficulties in obtaining reproducible structural foam mouldings of uniform surface appearance was that the polyol blends and conventional technical polymeric MDI used were only marginally compatible. In recent years this difficulty has been substantially overcome and the scope for polyol formulation widened by the development of several types of MDI variants such as Bayer's Desmodur 44V10, and ICI's Suprasec VM60. These isocyanates have a functionality of approximately 2·6 against the *ca* 2·8 functionality of the conventional MDI. Variants with functionality of *ca* 2·3 show even better compatibility, but lead to a substantial reduction in heat deflection temperature of the derived plastic.

In summary, there has been in recent years a significant improvement in processability and end properties of the system available for the production of structural urethane rigid foams and these, together with significant improvements in processing equipment, should ensure a substantial future market for these products.

The advantages of the polyurethane structural foam process can be summarised as follows:

1. A high strength to weight relationship in terms of density.
2. The ability to vary the density, and therefore physical properties, over a wide range, achieving economic advantages.
3. Due to excellent flow properties, the ability to produce larger and more complex components than by injection moulding processes.
4. Thickness variations from 5 to 100 mm produced in the same mould without sink marks.
5. The accurate reproduction of smooth or detailed surfaces, e.g. wood and leather grain effects.
6. The ability to mould *in-situ* inserts such as metal or wood.
7. A good balance of physical properties.
8. Low conversion costs.
9. The possibility to operate, sequentially, a number of moulds from a single machine delivering shot weights within a range of 0·5 to 25 kg, whereby production quantities of 1000 to 100 000 and above can be achieved.
10. Prototype and economic low volume production achieved with low cost moulds.

The use in Europe of all types of structural rigid foams in furniture has not grown as expected a few years ago, probably because furniture designs using these materials have not developed as expected. Uses are, however, developing for urethane structural foams in which the processability and technical end properties are utilised. Typical of such applications are window frames, lightweight fan blades for specialised machinery, etc.

Fibre Reinforced Structural Rigid Foams

The advantages to be obtained by the incorporation of steel and wooden struts, fibres, etc., have been recognised from the early stages of the development of urethane technology. Use has been made of the ability of the liquid reacting mix to foam around and adhere to a great variety of shapes and surfaces. Thus fabric reinforcement has been introduced into the surface of moulded flexible cushions and many modern refrigerator designs depend upon the ability of the rigid foam mix to act not only as the insulant but as the 'adhesive' locking together and stiffening the framework and metal sheets which comprise the refrigerator case.

Some continuous lamination and spray processes have used the concept of the simultaneous use of a chopper for glass rovings and a foam dispenser-spray unit.

In the specific field of structural foams, composites of wood and foam have been developed and manufactured. Wire reinforcement techniques have also been used to give significant increases in stiffness.[13]

Hoppe of Bayer published several patents on the 'Depotmat' process[14] which included the use of a reinforcement mat consisting of several layers of synthetic fibres designed so that as the foam expands it meets increasing resistance as it progresses to the outer skins. This gives an increasing foam density near the outer skins but retains the advantages of a low density core. This construction eliminated the high stress concentration at the discrete boundary of the outer skin and foam in normal foam laminates.

In conventional thermoplastic technology it has been known for some years that significant improvements in physical and mechanical properties can be achieved by reinforcement with glass fibres.[15]

The flexural modulus of thermoplastics such as polystyrene and polypropylene can be increased 2 to 3 times by the incorporation of about 15 % of glass fibre. Heat distortion temperature is raised, tensile strength is substantially increased and mould shrinkage reduced.

During the past two years an increasing amount of interest has been shown on the subject of fibre reinforced reaction injection moulded urethanes. The main driving force for this work has been the interest of General Motors and Ford USA in the use of moulded urethane elastomeric parts for the production of front and rear ends of autos (see Chapters 4 and 5). The early materials used in the auto field had low stiffness and high impact resistance but there is now a growing interest in the use of high modulus materials for doors, boot (trunk) lids, etc., and this has focused attention on the properties achieved by fibre reinforcement.

Mainly because of this interest of the auto companies, equipment manufacturers are now allocating a great deal of effort to the design of equipment suitable for the RIM materials incorporating glass fibre. The availability of equipment suitable for the automated production of glass-fibre reinforced structural rigid foams should open up a number of market outlets so far denied to structural foams.

The growth of usage can only be assessed when the new equipment has been used in production and economics and end properties established. In view, however, of the potential importance of fibre-reinforced structural foam and RIM the following summary of the present position is relevant. It must be emphasised that the speed of development in this field as a whole is such that it is difficult at present to predict what processes and outlets will eventually emerge.

Fibre Type

Two types of glass-fibre reinforcement are of interest—chopped strands and milled fibres.

Chopped strands are bundles of glass filaments which are mechanically cut to discrete lengths—typically $\frac{1}{8}$ in to $\frac{1}{4}$ in. Milled fibres are continuous glass filaments which have been hammer-milled into short lengths. They are designated by screen size, e.g. $\frac{1}{32}$ in, $\frac{1}{16}$ in, etc.

Development has mainly progressed with milled fibres because they can be more easily processed in prototype RIM equipment, since the viscosity of the dispersion is lower with milled fibres. Also, since they can be completely dispersed, smooth surfaces can be obtained on moulded parts. Viscosity is, of course, a critical parameter since most RIM processes depend upon high pressure impingement mixing. Excellent reviews of the present knowledge of the effect of glass-fibre type and size on the viscosity of the RIM components have been given by Isham.[16,17]

Equipment

RIM equipment is discussed in Chapter 9. It is at present believed that metering equipment for glass-fibre reinforced RIM should be based on high pressure metering cylinders since rapid and excessive wear appears to occur when using the high pressure axial piston pumps often used to meter conventional RIM materials. Isham[17] has reported that a combination of impingements and mechanical mixing gives excellent uniformity up to stream viscosities of 3000 cP.

More recently the development by Krauss-Maffei of equipment capable of the uniform dispersion of fibrous and particulate fillers in the RIM process has been reported.[18] Filled polyol is pumped to an accumulator cylinder. The amount of filled polyol required for each shot is forced out of this accumulator cylinder by a hydraulic piston into the lines to the high pressure head. Polyol/filler mixtures with viscosities up to 50 000 cP have been processed satisfactorily compared to 2000 cP on conventional RIM equipment. 50 % glass in polyol has been processed in this equipment.

Properties

Isham[17] has pointed out that the effect of glass-fibre reinforcement of urethanes can be divided between the improved properties achievable only with reinforcement and those influenced by the combination of glass fibres and urethane chemistry.

The unique glass-fibre reinforcement effect leads to a reduction of polymerisation shrinkage, reduction of the coefficient of thermal expansion and improved resistance to droop and sag at elevated temperatures. The reduction in the coefficient of thermal expansion is particularly important since this means that composites of metal and urethane can be designed with close tolerances between metal and urethane component. Resistance to sag at elevated temperatures is, of course, of importance when considering the ability of composites to withstand paint oven baking temperatures. The combination of glass fibre and urethane technology influences stiffness, tensile strength and tensile elongation. Over the range 0–25 %, mechanical

TABLE 1

RIM type	Thermal coefficient of expansion, 10^{-6} in/in/°F	Flexural modulus, psi	Tensile properties Strength, psi	Elongation, %
Flexible				
(a) Unreinforced	88	29	2 819	123
(b) 15 % Glass fibre	48	59	2 604	63
Semi-rigid				
(a) Unreinforced	71	121	4 508	83
(b) 15 % Glass fibre	36	213	4 905	27
Rigid				
(a) Unreinforced	60	260	7 691	16
(b) 15 % Glass fibre	35	430	8 722	11

properties generally show a linear relationship to the amount of reinforcement.

Properties achievable with glass-fibre reinforced rigid structural foams will depend upon the size, percentage and surface treatment of the fibre used and on the chemistry of the rigid urethane system used. Table 1 gives figures quoted by Isham[16] for the effect of 15 % of glass content. They are an indication of the level of improvement which can be expected. For completeness, Isham's figures for the effect on flexible, semi-rigid and rigid RIM urethanes are included. Isham defines these three categories as follows:

Flexible: RIM types used in flexible auto fascias, with flexural modulus of 20 000 psi to 40 000 psi.

Semi-rigid: RIM types proposed for auto fenders, etc. with modulus 100 000 to 150 000 psi.

Rigid: Flexural modulus 250 000 to 300 000 psi.

No data appear to have yet been published on the effect of up to 20 % milled glass-fibre content on the performance of structural rigid foams in fire tests but it is possible that improved fire performance may result. If this proves to be the case the market potential for structural rigid foams could well be substantially increased.

Morphology

In conventional composites the interface between the reinforcing agent and matrix plays an important role in controlling physical properties including the fracture process. Little has yet been published on this subject in relation to structural rigid foam.

Shortall and colleagues at Liverpool University are one group studying this phenomenon and have already published papers[19,20] on the morphology and fracture behaviour of polyurethane rigid foams reinforced by short fibres, and have correlated changes in morphology with change of tensile strength. For the foams studied they defined a critical fibre length for reinforcement of the foam systems and research of this type should enable optimisation of the properties of fibre-reinforced RIM to be achieved.

THE BACKING OF CARPETS WITH URETHANE FOAMS

Over recent years a number of methods of utilising flexible foam as carpet underlay and backing have been explored. These include bonded foam crumb, doctor knife techniques for laying down prefoamed mixes, the application of foamed layers by transfer techniques and direct spray laydown.

The use of urethane flexible foams as a foam backing for carpets was envisaged during the early days of the urethane industry. Smith and Wood[21] patented the foam backing of carpets by the spray application of a flexible urethane foam in 1958. At that time, however, knowledge of the chemistry of urethanes was not sufficiently advanced to allow low density foams with low levels of compression set, etc. to be obtained in thin layers and the process technology available was not suitable for laying down consistently thin layers of liquid urethane mixes. During recent years there has been a significant change in the status of urethane foams for the carpet backing industry and several types of urethane process are now at, or almost at, the stage of commercial exploitation. The interest in urethane foam backing has undoubtedly increased in recent years because of a growing realisation that, in addition to excellent properties, urethanes need a far lower energy input to achieve a cure and that because of the reduced energy usage a substantially reduced floor space occupation is possible.

Types of Process used for Urethane Foam Backing
Several proprietary processes have been developed for the application of urethane foams to carpets. They can, however, be divided into the following types:

Type 1. The use of existing latex coating equipment and techniques which involve the laydown of a slow reacting urethane mix by means of doctor blades. In some variants the urethane foam mix is mechanically frothed before application to allow control of the penetration of the mix.

Type 2. Transfer lamination techniques are used in which the urethane mix is laid down on a release paper or belt and is then transferred to the carpet when it has reached a tacky, partially reacted state.

Type 3. The direct laydown of a fast reacting urethane mix directly onto the carpet by a spray technique.

Type 1 Processes
This type of process has been developed and reported by several companies including Union Carbide[22] and Dow.[23] The advantages claimed for this method are that conventional equipment (sometimes slightly modified) can be used and that since the process is similar to existing latex processes no new skills are needed from the operators. The process is claimed to eliminate the need for a preliminary locking coat.

Cravens[23] has recently described a process of this type developed by Dow Chemical Company. The urethane foam mix is mechanically frothed and its rate of reaction controlled so that it can be doctored onto the carpet with conventional coating equipment. By moving the coated carpet through the oven the components react at the right time to control yarn penetration and to effect a full cure prior to exit. Density control over a wide range from about 80 lb/ft^3 to 12–15 lb/ft^3 is claimed.

Cravens has pointed out that significant labour cost reduction is possible by running at coating speeds of over 22 ft/min and increased yardage can be produced without the addition of extra oven investment. Energy consumption is significantly less than latex since urethane is a reactive system with no water or solvent removal necessary for cure. Relatively low temperatures are required for cure. Cravens' actual figures for relative gas consumption on a production line are reproduced in Table 2.

TABLE 2
RELATIVE GAS CONSUMPTION
(Actual Measurements on Production Line)

	Cubic feet gas/square yard
Latex fat foam and latex precoat 42 oz/yd^2 + 20 oz/yd^2 Speed—6·5 fpm	13·8
Latex foam and latex precoat 24 oz/yd^2 + 20 oz/yd^2 Speed—12 fpm	9·8
Jute lamination with latex 30 oz/yd^2 Speed—18 fpm	4·9
Urethane sponge 56 oz/yd^2 Speed—22 fpm	2·8

As with the other types of urethane process, the lower temperatures associated with urethane curing means improved working conditions around the oven and if a polypropylene backing is in use with the carpet no melting of the backing occurs because of the lower cure temperature used.

The advantages of mechanical frothing and knife coating using essentially existing equipment and skills are obvious but as yet low densities have not been achieved, probably because of the need for a slow reacting mix.

Type 2 Transfer Processes

One process of this type is the Bayer/Metzeler process. The Bayer/Metzeler process is claimed[24,25] to foam back carpets up to 5 m in width at a speed of up to 6 m/min. The urethane reaction mix is applied by a traverse and spread as a thin layer by doctor roll on a specially designed large diameter drum held at 80 °C. Shortly afterwards the carpet comes into contact with the reaction mix. As the drum rotates the urethane mix combines with the

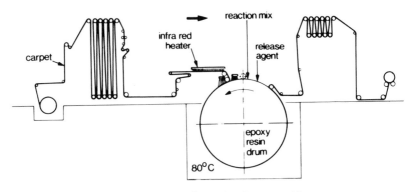

FIG. 1. The Bayer/Metzeler drum machine.

carpet fabric and at the same time expands. After one revolution the reaction is complete, and the backed carpet is drawn off. A drawing of the process is shown in Fig. 1. It is claimed that the use of a drum enables several narrow widths of carpet to be backed side by side.

The transfer type process allows low densities to be achieved and foam faults to be identified before contact with the carpet. It is also possible to control penetration of the urethane mix into the carpet by adjusting the formulation, cure and/or time of transfer. This might make it easier to obtain uniform penetration of a variety of carpet types.

The process tends to be more capital intensive, however, and usually involves the additional cost of the transfer paper used. A locking coat appears also to be necessary before application of the foam.

Type 3 Process—Direct Laydown

The process developed by ICI uses direct spray application of the liquid urethane mix to the carpet back (Fig. 2). Advantages in cost, carpet stability and versatility are claimed.[26] ICI argue that the lowest costs in capital, raw material usage and heating costs are achieved if a reactive urethane mix,

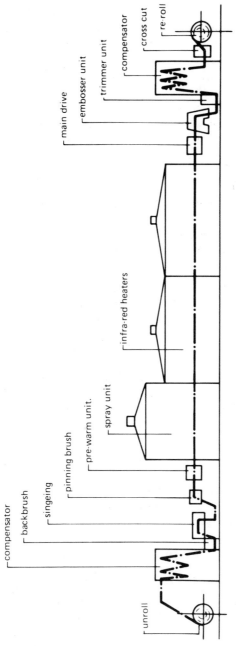

FIG. 2.   Typical layout arrangement of ICI carpet backing machine.

filled or unfilled, is laid down directly onto the back of the carpet without the use of a preliminary adhesive/tuft lock stage and without the use of expensive transfer techniques/release agents. In a typical application of a 5 mm foam back, about 30% of the applied reacting mix penetrates the tufts and primary backing giving excellent dimensional stability, tuft lock and fibre lock, while the remaining mix reacts to form the foam back.

Consideration of optimum design requirements for 5 m wide machines operating at 6 m/min defined a twin spray machine with a total output of approximately 24 kg/min and traverse speeds in excess of 200 m/min. Modification of the traverse design of the ICI/Viking continuous laminating machines with their electromagnetic clutch turn round is expected to enable excellent uniformity of laydown to be achieved. Traverse performance is, of course, quite critical in this application.

Some carpet backs have loose ends and extraneous fibres which will cause gross defects in the foam back when using direct application techniques. These are prevented by singeing the carpet and a singeing unit is fitted to the carpet feed side of the machine.

The surface quality obtained by spray laydown is technically satisfactory having thickness variation less than 0·05 mm but the surface still does not have the matt uniform surface of a knife coated latex or PVC. Heavy embossing is not acceptable to the majority of European potential customers but a light emboss to give a modified surface pattern has been demonstrated and found acceptable. The original concept was to emboss the foam immediately after rise and as soon as the surface was tack free. This method, however, suffers from the disadvantage of needing accurate timing and positioning of the embossing roll with the need for adjustment as the production proceeded. It has now been demonstrated, however, that a satisfactory kiss embossing can be applied in a later stage of the process by using a heated embossing roll. The embossing roll might also be used to contribute to the control of foam thickness.

At the time of writing, the ICI direct application process has still to be proven in continuous multi-shift operation on a 5 m wide line.

As mentioned above, the mechanical requirements on the traverse are critical and optimisation of design will be necessary. Further experience of the effect of variations in carpet weave and surface on the amount of foam penetration is also necessary. Data obtained in 4 h continuous runs indicate, however, that uniform low density urethane foam backing is feasible by direct spray. Very little heat curing appears to be necessary. Coating weights down to 650 g/m^2 have been produced and thickness variations of only 0·05 mm achieved on backings of 3 to 5 mm thickness.

Urethane Processes versus Latex Backing

In summary, most of the urethane processes developed claim the following advantages over the latex foam process:

1. lower capital cost;
2. lower labour requirements;
3. lower energy requirements;
4. most avoid the tuft locking coat;
5. less space required; and
6. lower weight of finished carpets.

Physical Properties of Urethane Foams used as Carpet Backing

It is generally accepted that the physical properties of polyurethane foams are superior with respect to most wear properties than conventional latex or PVC foams.

Properties quoted by Cravens[23] for $12-15\,lb/ft^3$ foam applied by the Dow process are given in Table 3.

TABLE 3

PHYSICAL PROPERTIES OF URETHANE SPONGE BACKED CARPET

Property	Typical property range
Density (lb/ft^3)	12–15
Tensile strength (psi)	47–55
Elongation (%)	130–150
Tear strength (lb)	3–5
Compression set (%)	3–10
RMA modulus (psi)	9–12
Delamination (lb/in)	2–5
Tuft lock (lb)[a]	6–18
Resiliency (ball bounce %)	38–44
Fibre bundle penetration (%)	100
Pilling and fuzzing resistance	Excellent
Water resistance	Excellent
Accelerated ageing	
Properties	Excellent
Colour	Darkens slightly

[a] Depends on carpet yarn and backing construction.

Bobe, Hurd and Woods[26] have reported on some of the properties of carpets backed with lower density urethane foams produced by the ICI direct laydown method.

The physical property advantages of urethane foam backings can be summarised as follows:

1. Good abrasion resistance.
2. Excellent fibre-lock and resistance to fuzzing.
3. Good resistance to transit and handling damage.
4. Improved carpet dimensional stability.
5. Good moisture resistance.
6. Good 'castor' resistance.
7. Good thermal insulation.
8. High permeability.
9. Good heat ageing performance.
10. Light-weight, easier to lay.

Comfort Factor
The 'walking characteristics' of carpets can only be judged subjectively and have much in common with the problem of judging comfort in seating cushions. ICI state that service trials which involved questioning the people using the carpets in a crowded office environment revealed that the polyurethane backed carpets were preferred to soft latex foam backed carpets. It seems that most people do not like their foot to fall rapidly until 'bottoming' on the hard floor. This difference was characterised in behaviour by a stepwise loading curve using the Wool Industries Research Association (WIRA) gauge. The curve for a polyurethane backing and for a typical average weight latex backing is shown in Fig. 3. The lower rate of descent and the lower penetration at 12 psi load of the polyurethane are indicative of the better walking characteristics described.

Economics of the Process
Any comparison of cost between urethane and latex foam backing is extremely difficult because of the wide variation which can be made in the formulation and density of a rubber latex foam. Latex foam can be made using increasing quantities of filler to give a cheaper foam but the addition of filler degrades the quality of the foam. Reduction in density of a rubber latex foam also gives inferior physical properties but these may be adequate for many types of application.

Some years ago the cost of latex mixes per pound were substantially below those of urethanes and, with cheap energy available, urethane backing could only be justified for higher quality carpets. There is now less difference in raw material costs and substantially increased labour and energy costs have favoured urethane processes.

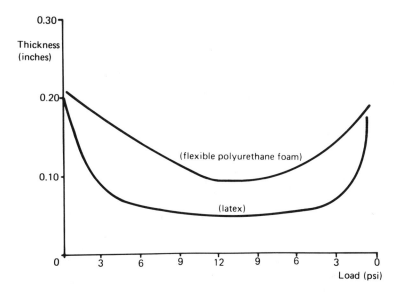

FIG. 3.   Comparison of the walking characteristics of flexible polyurethane foam carpet backing and latex carpet backing when submitted to WIRA carpet gauge measurements.

Some cost comparisons of the Bayer/Metzeler process and latex backing have been published.[25] These show in favour of urethanes a lower investment, and a third of the annual cost of the space required. Direct staff costs are less (8 men for the latex line, 5 men for the urethane line). Substantial savings on electricity and on heat (£44 000/year) are also claimed.

A direct application process might offer further advantages since it eliminates the need for the use of release agents or release papers (approximately 4p/yd^2) and for a latex precoat (this can cost up to 12p/yd^2).

Size of Market
The total Western European production of tufted carpets in 1976 has been estimated as approximately 430 million square metres. Of this large area of carpet produced only a proportion has a foam backing and this depends upon the type of carpet being produced and the fashion trends in the country concerned. The normal higher quality woven carpets are not usually covered by a foam backing and it is only the tufted and needlefelt carpets that require foam backing to give the carpet improved dimensional

stability, increased wearing properties, more attractive appearance and increased comfort.

Estimates (1976) suggest that 78 % of all UK tufted carpets were foam backed in 1975, 87 % of all European tufted carpets but only 30 % of the USA carpet production. Taking the European figures of production this would indicate that, at 900 g/m² of foam backing, approximately 230 000 tonnes of backing was used: mainly natural and synthetic rubber latex.

Assuming a modest increase of 5 to 6 % per year, production in Western Europe in 1980 will be of the order of 535 million square metres. Of this production it is anticipated that 90 %, approximately 480 million square metres, will be foam backed and if by 1980 10 % of this foam backing is urethane this will represent a potential market of 30 000 tonnes of urethane chemicals. In view of the availability of the various processes referred to above this seems realistic.

The penetration of the latex backing market by urethane will probably be controlled by the rate at which existing latex machines are modified to use urethanes (a relatively cheap operation costing approximately £50 000 per machine), in addition to the installation of new lines using urethane only.

The Fire Resistance of Carpet Underlays

In 1969 the US Government recognised the need for a flammability standard for carpets and rugs.[27] This led to the first US National Standard[28] which was a pile test designed to reduce the probability of carpet ignition.

It was subsequently recognised, however, that with some cases, such as in the corridors of institutional buildings, multiple occupancy buildings, etc., a more severe test of the fire resistance of floor coverings is desirable. Studies were made of the flame spread in a corridor of various types of floor covering and this test is now referred to as the National Bureau of Standards Corridor test.

In 1973 and 1974 the Urethane Safety Group of the Society of the Plastics Industry sponsored a series of large-scale fire tests at the National Bureau of Standards to evaluate the role of various types of underlayments on the fire performance of a carpet installed in a corridor.

Five different types of underlayment were used in the testing programme—hair jute, virgin and rebonded urethane, integral latex and a styrene butadiene pad. It was found that the insulating effect of all the underlayments lowered the critical radiant flux of the carpet systems tested so that the carpets burned a greater distance compared to the carpet alone. No significant differences were found in the performance of the various underlayments. The Urethanes Safety Group of SPI have published a bulletin giving details on these findings.[29] The National Bureau of Standards has published data on the development of criteria for the use of a small-scale critical radiant flux test method for assessing the potential flame spread of flooring systems.[30]

REFERENCES

1. German Patents 851,100, 853,438, 864,917, and others.
2. *Polymer Composites of Wood and Agricultural Residues Europe, North America, Japan*, 1973–1980, Pub. De Bell & Richardson Inc.
3. DEPPE, H. J. (1977). Federal Institute for Materials Testing, Berlin, Paper to Particle Board Symposium, Washington State University, 22 Mar.
4. Belgian Patent 839,546. A. M. Wooler to ICI Ltd.
5. ICI UK Patent 847,127.
6. ICI US Patent 3,830,760.
7. TAMURA, G. T. Measurement of Air Leakage Characteristics of House Enclosures, Building Services Section, N.R.C., Ottawa, Canada.
8. CLEGG, W. R. (1976). In *Plastics in Furniture*, D. M. Buttrey (Ed). Applied Science, London, p. 85.
9. HILL, A. C. (1969). *Bulletin, Institute of Body Engineers*, May, 35.
10. HILL, A. C. (1972). *Engineering Materials and Design*, Apr.
11. *Modern Plastics International*, 1972, **2**(6), June, 14.
12. JACOBS, K. (1977). *J. Cell. Plast.*, **13**(2), Mar./Apr., 133.
13. *Modern Plastics International*, 1976, **6**(7), July, 8.
14. HOPPE, P. UK Patent 1,191,902 (1970) and UK Patent 1,233,910 (1971).
15. *Material Engineering*, 1971, **3**(5), 33.
16. ISHAM, A. B. (1977). Paper to the International Conference of Polymer Processing MIT, 15–18 Aug.
17. ISHAM, A. B. (1976). Society of Automotive Engineers Congress, Detroit. 23–27 Feb.
18. *Modern Plastics International*, 1977, **7**(9), Sept., 14.
19. COTGREAVE, T. and SHORTALL, J. B. (1977). *J. Mat. Sci.*, **12**, 708.
20. SHORTALL, J. B. (1976). *Proc. 4th SPI International Cellular Plastics Conference*, Montreal, Mar., 134.
21. SMITH, W. and WOOD, J. F. W. UK Patent 847,127 (16.1.58).
22. MARLIN, L., DURANTE, A. J. and SCHWARZ, E. (1975). *J. Cell. Plast.*, **11**, 6, Nov./Dec., 317.
23. CRAVENS, T. E. (1976). *Carpet and Rug Industry*, Oct.
24. *Plastics and Rubber Weekly*, 1975, 7th Mar., 25; 1974, Oct., 859.
25. Chemlefasern/Textil-Industrie, 1975, Jan., 48.
26. BOBE, J., HURD, R. and WOODS, G. (1976). *Proc. 4th International SPI Conference*, Montreal, Nov., 290.
27. US Federal Register (34 FR 19812), 1969, 18 Dec.

28. Standard for the Surface Flammability of Carpets and Rugs, US National Standard DOC FF 1-70, 1970.
29. SPI U.S.G. Bulletin U105, 'The Evaluation of the Fire Performance of Carpet Underlayments'.
30. BENJAMIN, I. A. and ADAMS, H. C. (1975). National Bureau of Standards Report, NBSIR 75-950, Dec.

Chapter 9

DEVELOPMENTS IN POLYURETHANE MACHINERY

J. B. BLACKWELL

Polymech Ltd, Stockport, UK

and

R. RUBATTO

Studio Tecnico Mazzucco–Rubatto, Turin, Italy

SUMMARY

This chapter is concerned with machinery developments in the field of foam urethane application technology. In flexible foam slabstock the importance of plant size is examined together with the economy in using the various systems for flat top production. The continuous manufacture of rigid foam laminates using floating platen and inverse laminator techniques is discussed. The improvements in hot-cure and cold-cure moulding lines are illustrated, together with the benefits of reduced labour costs by employing a high degree of automation on the production line. Finally, the equipment for Reaction Injection Moulding (RIM) in the manufacture of a wide range of automotive and structural foam components is discussed.

INTRODUCTION

Probably the most significant difference between the manufacture of expanded polyurethane process and any other plastic is that, in general terms, the polymer is made for the first time in the end users' plant or

223

factory, rather than in the factory of the raw material supplier. This leads to a number of engineering (apart from the chemicals and process) problems. The known requirements of accurate metering, temperature control and good mixing to produce the polymer, linked to conveying, moulding and subsequent storage have been added to by the considerable rate of growth and development of the processes. Further, the relative value of the chemicals to the cost of the machinery is disproportionate when compared with other plastic processes. This chapter will deal with developments in flexible slabstock machinery, rigid foam laminator machinery, automobile trim moulding machinery, and reaction injection moulding equipment.

FLEXIBLE SLABSTOCK MACHINERY

During the early 1960s flexible slabstock equipment was developing in terms of mixing head throughput in kilogrammes per minute and increased size and number of machines. Machinery, which was suitable for polyester foams, was changed to meet the needs of polyether foams, initially through prepolymers and then by the 'one-shot systems'. The problems of handling the higher viscosity polyester resins were exchanged for the increased output and increased number of components of the polyether systems. In 1965, machine outputs of 250 kg/min were being introduced and Table 1 shows a typical output and number of components for a one-shot polyether plant. This point in time also saw the firm establishment of the low pressure system as the preferred method of making the polyether type of flexible foams. An important problem was the reduction of skin waste and this led to the demand of further increased output: since the smaller the block the greater the percentage skin loss.

Plants with outputs of 500 kg/min were designed and manufactured to produce blocks 2 m wide by 1·2 m high at densities from 16 kg/m³ to 30 kg/m³. Higher densities can be made but with a diminution in block height. A typical 500 kg/min plant is shown in Fig. 1. Some plants were combined high and low pressure units, with high pressure for polyester product and low pressure for polyether product. These higher outputs produced the following machinery changes.

(a) Metering Units
Increased polyol pump output was obtained by choosing larger size gear pumps, but in the case of TDI initially larger pumps were not available, and multi-pump units were used. As the number of pumps increased this raised

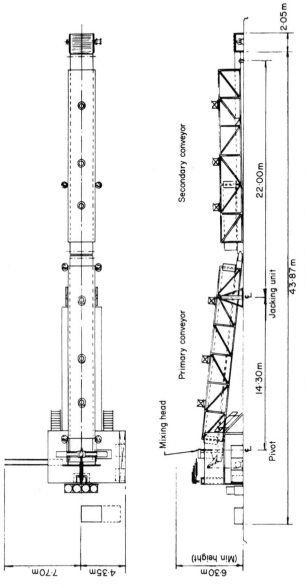

FIG. 1. Flexible slabstock plant, 500 kg/min: third angle projection.

TABLE 1

Component	Output range kg/min
Polyether	43–160
TDI 80/20	15–60
Amine catalyst	0·2–1·6
Silicon	0·7–6·8
Water	1·4–10
Tin catalyst	0·1–0·9
CMF 11	2·0–16·0

further mechanical problems and, eventually, in both high and low pressure applications a single specially modified hydraulic pump of swash plate design was chosen. Other types of pumps can be used but accuracy of metering—when operating under varying back pressure conditions especially with low viscosity liquids—is suspect. The increase in power, together with variable output, led to the use of infinitely variable speed drive units of the hydraulic or preferably electronically controlled eddy current coupling types; such as Tasc units[1] which can be easily remotely controlled.

(b) Mixing Head
As throughput changed it became necessary to increase the dimensions of the mixing chamber and, consequently, the power of the mixing head drive. The heavy reciprocating weight on the traverse led to the introduction of a hydraulic mixing head drive which reduced the traverse weight considerably. For example, a 50 horsepower variable speed DC motor weighs 500 kg against a hydraulic motor of 22·5 kg. A typical hydraulic drive unit, which is floor mounted, is shown in Fig. 2. The exit velocity from the mixing head is also important, as too high a velocity can cause splashing on the paper side walls and if the nozzle is facing down stream this high velocity can cause under-running and splits. This led to the development of the 'Honeycomb Nozzle'[2] in which the flow from the mixing chamber passes through a 'Venturi Nozzle' and then through the Honeycomb section, where the exit velocity is considerably reduced and allows the mixed liquids to be laid down uniformly without splashing. This device also permits the conveyor to be run at a slightly lower speed, enabling high blocks to be produced. Many types of traverse mechanisms have been tried—pneumatic, hydraulic—but the most popular type is the electro-magnetic unit where the turn-round speed can be adjusted electrically at the

end of the traverse to compensate for the acceleration and deceleration at turn round thereby reducing build-up at the shoulders of the foam block.

(c) *Conveyors*

It is normal practice to pour flexible foam on to a sloping conveyor where the angle can be changed from 2° to 5°. The process chemistry establishes

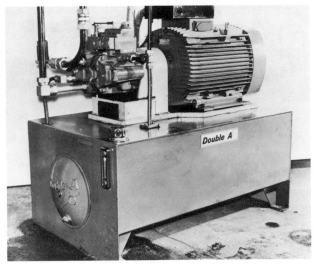

FIG. 2. Hydraulic power pack for mixer head drive for high output slabstock plants.

the conveyor length in terms of (a) the initial conveyor including side walls, (b) the secondary conveyor without side walls used to transport the foam or block until it is cut into length, prior to storage. Again, higher throughputs give higher blocks which, in turn, means longer conveyors with higher side walls and higher speeds of operation. The old original belt conveyors have been largely superseded by metal slat conveyors which are easier to clean and do not have the difficulty of belt tracking which can cause paper problems. The cream time, full rise time and time before the blocks are cut, have led to primary slat conveyors up to 26 m long and the secondary conveyor 14 m long, suitable for processing slabstock urethane blocks at 500 kg/min (see Fig. 3). The height to the operator's platform is 3·8 m above the ground. Infinitely variable speed of the drive with remote control from the operator's platform is essential on modern high output installations.

FIG. 3.   High output slabstock—variable fulcrum support at entry end of primary
conveyor.

(d) Pipework

The urethane manufacturing process is sensitive to air—it can, for example,
upset metering if air is present—and the pumps are normally volumetric
delivery units. Air also acts as a nucleating point around which the foam
grows and can affect the cell size. In low pressure machines controlled
amounts of air are injected into the mixing head and this, in conjunction
with mixer speed variation, can be used to change the cell structure. In high
pressure machines cell structure is controlled by means of back pressure
changes at the mixing head exit nozzle. This pressure drop in the mixing
chamber releases the air, which is dissolved—primarily in the TDI. Thus
random amounts of air are undesirable. Attention to pipe diameter sizing,
avoidance of air pockets in pipe layout, keeping the inlet lines to the pumps
as short as possible, are important features of piping design. Too high a
back pressure on a low viscosity stream where the pump is sensitive to back
pressure change can lead to irregular metering and consequent foam faults.
Long inlet pipes can cause cavitation of the pump and again produce
irregular metering.

(e) Reproducibility and Temperature Control—Automatic Operation

The cream line on a foam slabstock plant is the line where the mixed liquids
start to gradually increase in viscosity and expand. If the distance from the
mixing head to the cream line could be kept constant, then it should lead to

constant and reproducible block production. The factors influencing this, however—ignoring chemical variation—are:

(a) Precise temperature control of all the reactants, but essentially the major components, polyol and TDI.
(b) Accuracy and reproducibility of metering.
(c) Uniform conveyor speed.
(d) Temperature control in the area of the foam producing plant until the foam is fully risen.
(e) Uniform traverse speed.
(f) Constant mixing head speed.

In addition, modern high throughput installations have the requirement to change colour, throughput and formulation without stopping the plant, as

FIG. 4.   Metering unit control panel high output slabstock plant.

each stop and start means foam losses. The financial penalties for losses, due to faulty start-up and shutdown—not only in raw materials but in fire hazard—have led to the introduction of automatic sequence control of start-up and shutdown. At 500 kg/min the cost can be high without these refinements. They also take the responsibility from the operator of switching the chemical streams on and off in the correct order. Figure 4 shows a typical control panel for a high output slabstock plant.

TABLE 2
AVERAGE SKIN LOSSES CROWN BLOCK

Skin	Percentage
Top	7–8
Bottom	4–5
Side	2–3
Total	13–16

Loss Reduction in Slabstock Production

From an examination of the cross-section of a block of foam an analysis of
the skin losses are shown in Table 2. The initial work was to make the block
of foam square, i.e. reducing the dome or crown on the block, and is known
as the Draka Square Block Process.[3] One effect that produces the crown is
the frictional drag of the foam on the side paper during the expansion
process. Methods were introduced whereby polythene sheets were fed in
between the paper sides and the foam and these sheets were pulled upwards
for a proportion of the rise time, normally starting at about 30 % of the rise
profile and finishing about 75–80 % of full rise. Figure 5 shows the effect of

FIG. 5.   Flat topped block foam.

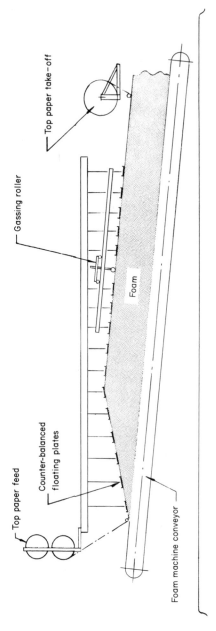

FIG. 6. 'Planiblock' method.

pulling the foam on one side, the other being left to free rise. This process has now been in commercial operation since 1967 and has resulted in a worldwide licensing of the method.

Planiblock Process[4]

In 1973 an alternative process was developed in Spain. By this method a paper is placed across the top of the foam during the foam rise and is kept in position by a series of counter-balanced platens which flatten the top of the block. Yields up to 94 % of prime foam are claimed for this process. Figure 6 shows a line diagram of the method involved. Another method of restrained rise profile was developed in Germany.[5]

Maxfoam Process[6]

A new approach to slabstock manufacture was initiated by Laader Berg in 1970, which was subsequently developed by Unifoam—this is the Maxfoam process. The principle of operation is as shown in Fig. 7. The mixed

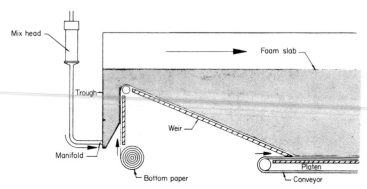

FIG. 7.   The 'Maxfoam' process.

urethane components are fed into the bottom of a metal trough. The liquids start to react and expand and the resultant foam liquid flows over the edge of the trough and is deposited on to the moving paper. The principal expansion takes place downwards over a sloping inclined pour plate, which deposits the foam on to a horizontal conveyor. The fact that the rising foam is led down an incline gives a flat top to the block. The process runs more slowly than conventional foaming and the conveyors are horizontal which means that the overall plant size—and output for a given block size—are reduced in comparison with conventional foaming. It was not possible to change the width without stopping until the recent development of the Varimax process.[7]

The Future

The above effects to reduce the waste from flexible foam slabstock production have produced important savings for the industry. It is equally certain that further developments will continue to be made—the size of the flexible slabstock industry alone will ensure this happens.

RIGID FOAM LAMINATOR MACHINERY

The manufacture of panels with a rigid urethane core has been established for some time, mainly through press injection. During the early 1960s work was carried out on a continuous form of lamination between various substrates and the principles of operation are well known.[8] Briefly, rigid urethane chemicals are accurately metered, mixed and laid uniformly across a substrate and then a top facing material is fed over the rising foam mass, after which the composite is then passed between an upper and lower conveyor. The two conveyors can either be of the fixed gap type or of the floating top platen type and where the foam to a large extent (95%) has a free rise before passing under the pneumatically controlled individual platens. This is illustrated in Fig. 8 and is known as the floating platen technique.

One function of the floating platen is to reduce the build-up and rolling of the foam at the entrance to the twin conveyor. This build-up tends to produce a poor cell structure at the top interface between foam and substrate leading to poor adhesion and reduction in physical properties. The key relationship is density to physical properties, i.e. to obtain the lowest density with maximum physical properties. After passing between the two conveyors the laminate can be side trimmed and cut to length. Laminators of this type run up to 10 m/min at thicknesses up to 12 cm.

Higher speed lamination has been made feasible (up to 50 m/min) using the principle of a doctor roll and free rise of foam.

These types of laminator are normally used with flexible facings, e.g. paper and bitumen felt, but further developments have led to the use of rigid facings.

Inverse Lamination

An extremely versatile laminator, which is suitable for the manufacture of

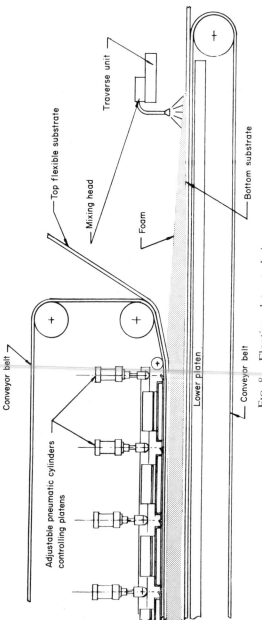

FIG. 8.   Floating platen technique.

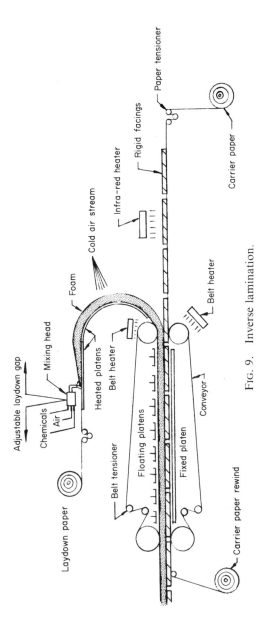

FIG. 9.   Inverse lamination.

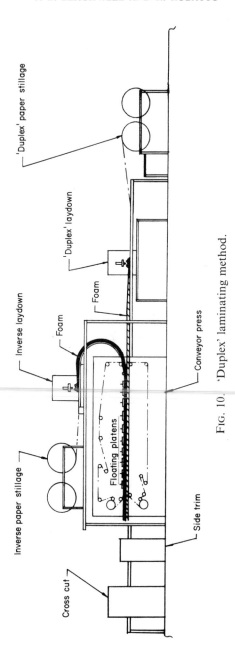

FIG. 10. 'Duplex' laminating method.

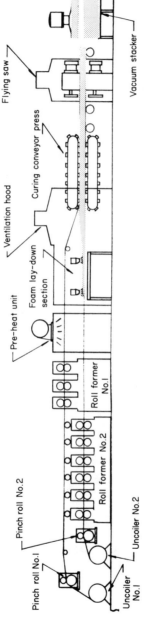

FIG. 11. Continuous rigid foam laminating line for metal faced sandwich panels.

flexible facing and one rigid facing, is of the Inverse type. The principle of operation is illustrated in Fig. 9 which shows the production of a flexible/rigid facing with a urethane core. The accurately metered and mixed chemicals are laid down uniformly on to a top moving paper which is then passed over a heated semi-circular platen. The foam rises as it passes over the platen and eventually meets the bottom substrate—in this case rigid sheets. The rigid sheets can be fed automatically by means of a vacuum transfer unit on to a carrier paper used as a method of movement of the rigid sheets. The composite then enters the double conveyor unit. The top conveyor is of the floating platen type. After passing through the double conveyor the bottom carrier paper is rewound and a sensor detects the gap between the rigid facings and this actuates the cross cut unit to produce discrete panels.

A further development of this process is the Duplex system, see Fig. 10. Basically, a second mixing head and traverse unit is added in the horizontal position. This enables thicker laminates in which both the facings are very flat and stable with a minimum edge trim loss. It is also possible to obtain double flexible faced laminates at twice the speed.

Continuous Double Metal Laminates

The use of metal faced rigid urethane foam panels has reduced the labour intensive steps of fabricating insulated walling on site. The process generally illustrated in Fig. 11 consists of taking metal facings, uncoiling and moving them through successive stages of metal roll forms before passing them through the foam distributor and double conveyor, after which the panel goes through a flying saw and on to a vacuum stacker unit. A single line can produce finished sandwich panels up to 1250 mm wide at production speeds up to 7 m/min. A variety of profiles can be formed and edge interlocks incorporated. When required, paper or felt facing can be applied on one side in place of the metal.

AUTOMOBILE TRIM MOULDING MACHINERY

Improvements of Hot Cure Moulding Lines

In the early years of polyurethane flexible foam moulding a two-step process, using prepolymers, was developed; with the use of stannous octoate catalyst and more active silicone surfactants, a one-shot process was introduced into moulding production. Metering units with several individual streams and closely controlled high-shear mixing heads were developed; at the same time different curing systems were used—convection hot-air oven, radiant heating and microwave radiation. This last system is not in large-scale use.

In the middle of 1970, General Motors introduced a new seating concept based on thicker automobile urethane foam seat pads. These thicker pads called 'full-depth' seats were used in many GM models in 1971; these new

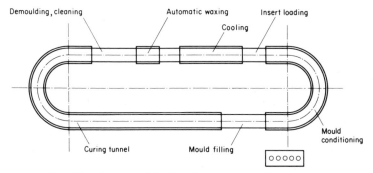

FIG. 12.   Layout of flexible foam hot-moulding plant.

cushions were simpler to assemble and reduced by 85 % the total number of seat parts; the moulded parts were made using one-shot TDI based formulations, and high temperature, radiant or high velocity, high temperature air ovens to cure the foam. The introduction of 'full-depth' cushions at the beginning of the 1970s encouraged improvements in the lines and decreased the cycle times. Flexible foam moulding lines today are highly automated with high production rates, and are capable of high versatility to mould a great variety of foam parts required by the market. At the pouring station the mixing head fills the mould following an automatic programmed pattern and also the required formulation has been programmed; normally in hot cure foam the filling is made in open moulds. The main improvements of these later years, apart from higher automation, are shorter cycle times: from 24 to 30 min down to 18 to 20 min; this improvement has been possible through new polyols with higher levels of primary OH groups. A typical layout of a hot cure flexible foam moulding line is illustrated in Fig. 12.

Cold Flexible Foam Moulding Developments
The introduction of 'full-depth' cushions requires higher 'SAG' factors, and

the specifications call for improved fire retardancy of the foam. The advantages of lower processing temperatures and shorter cure cycles increased the interest of the producers in the foam described as 'cold-moulding', 'cold-cure' or high resilience (HR). Another advantage of this type of foam is that it feels like latex, i.e. with high resilience. This foam is characterised in that the polyols used are more reactive; highly ethylene oxide tipped (usually greater than 50% primary OH content) with a molecular weight from 4500 to 6500. Initially, moulded cold foam was characterised by the use of a blend of crude MDI/pure TDI in the ratio 60/40 up to 20/80 or by using crude TDI; currently the moulded cold foam can also be produced with pure TDI. Initially the cold-moulded foam has been used in the furniture field, because of its latex-like feel and the possibility to foam into room temperature moulds without post-cure. In fact the term cold foam means foaming and curing at room temperature. This was possible with formulations based on MDI/TDI blends where such additives as triethanol amine and liquid aromatic amines were included to further increase reactivity and the exothermic reaction. Subsequently the advantages of cold foam with higher SAG factor, improved fire-resistance, better durability to fatigue, and shorter cycle times encouraged the automotive industry to use it for the production of 'full-depth' cushions. The introduction of cold high resilient urethane foam for automotive use has had a considerable influence in improving the production lines and the degree of automation in order to be able to produce many thousands of pieces per day in a reproducible manner.

Theoretically the processing of cold foam is inherently simpler than conventional hot moulding, because the cycle times can be reduced by about 50% and the moulded parts can be stripped in a reasonable time (about 10 min) without using external heat during the cure cycle. The foam can be produced successfully with different types of mould: aluminium, steel, expoxyfibre-glass, or integral rigid polyurethane foam. However, in order to have a repeatable production, today, the major plants have ovens for mould conditioning and the same type of mould must be used, preferably aluminium. Figure 13 illustrates a typical layout of a 'cold'-cure flexible moulding line, where the oven section for maintaining the reaction heat and for mould conditioning is about two-thirds of the line. The temperature must be such that the mould arrives at the pouring station at the required temperature, depending on the formulation requirements. Mould filling temperatures are normally 40 to 45 °C for the formulations based on MDI/TDI blend in the ratio 20/80 and about 50 °C for the formulation based on TDI. Some formulations can be injected in the mould at 55 or 60 °C. The oven conditioning temperature is between 60 and 80 °C. At the pouring station the mixing head must fill the mould following an automatic pattern because high resilient foam does not flow as freely as the conventional hot foam in the mould: the longest movement is about 10 cm. Other types of plant do not have a conditioning oven, but in this case all the moulds are conditioned at the required temperature optimal for filling; aluminium moulds with a thermostatic cavity and warm water being used.

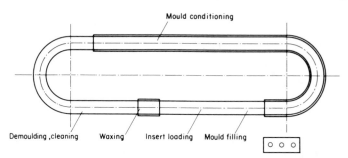

Fig. 13.   Layout of flexible foam cold-moulding plant.

If conditioned moulds are used, there is the possibility of having an independent heating unit connected to each mould, so permitting different mould temperatures at the injection point in the same plant; moreover, the programmed mould temperature is always the same. Normally conditioned moulds are used in small plants (up to 20 units approximately) or for large moulded parts such as chairs for the furniture industry. Big plants producing a high number of cushions for the car industry normally have an oven for mould conditioning.

To save labour costs, improvements in automation have been made: automatically closing and opening of the moulds by a cam action top lid rail, automatic programmed pattern in the mould with a formulation preselection possibility, and automatic release agent application. For this last purpose some sophisticated lines have used robots with programmed patterns for each mould. These have the possibility of applying a homogeneous and thin layer of release agent with the following advantages: saving of about 80 % of release agent and less cleaning of the moulds due to using lower quantities of release agent. The solvents from release agents are vaporised easily from the mould at 45 °C. When water-based release agents, which can be applied successfully in cold moulding, are used, a stream of hot air must be applied in the different points of the mould. Two-component metering units were used in the beginning of cold-foam moulding. Later metering units with 3–5 components have been developed. This is useful when different foam specifications apart from hardness are being met. When the variants in the specifications only cover different hardnesses a 3-component metering unit is used. Higher output machines than conventional have been introduced to overcome the poor flow of the foam.

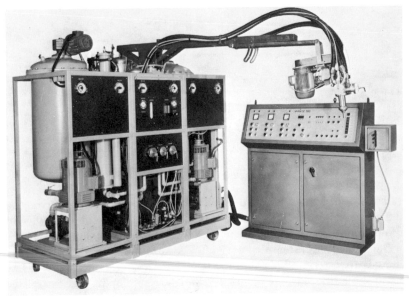

FIG. 14.   Automatic three-component unit for cold foam.

A standard 3-component machine for filling open moulds is illustrated in Fig. 14. The output is variable up to 200 kg/min with ratios infinitely variable from 1:1 to 10:1 with automatic preselection of pumps output using DC motors; programming is through high reliability proximity contactor cards. The three components are: (1) a blend of polyol with water, cross-linkers, surfactant and catalysts; (2) isocyanate or blend of isocyanate—the pump is driven by a DC motor which gives an index range from 85 to 115; (3) CMF 11 infinitely variable from zero to 20 parts referred to 100 parts of polyol blend.

For the filling in open or in closed moulds (through a hole of 50 mm diameter in the centre of the lid which is automatically closed after filling), high pressure metering units with 2 or 3 components can be used. In the field of high pressure machines for cold flexible foam, the development of the self-cleaning mixing head avoiding the necessity of any solvent flushing is important.

In the mould design, improvements were made in the closing profile. In the early 1970s, very heavy moulds with very strong clamps were used in plants to avoid foam flash from the horizontal closing profile due to the high pressure generated inside those moulds which did not have holes in the lid. Now the lid has small holes—2 to 3 mm diameter—which enable the air to escape, but a further improvement is that the closing profile is conical with different angles.

A new concept that is now developing is the static plant. Instead of being positioned on the carriage of the conveyor, the mould is placed between two platens of a press or a mould-holder with built-in clamping and the material is injected preferably by a high pressure machine with a self-cleaning mixing head in the closed thermostatic mould. This new system is possible because of new fast curing formulations with a cycle time in the mould less than 2 min.[9]

By elimination of high heat curing in cold moulding, urethane foam can be poured directly into vacuum formed plastic or textile covers to produce complex parts without deformation of the covering. This new system was developed in the 1970s for production of prototypes and seats of small production cars. An appreciable effort has also been made in moulding highly resilient flexible foam seats over a rigid plastic shell. The forecast is that the next advance in moulding technology will utilise processing advantages to produce one-shot production of automotive seating complete with cover and insert.

While current efforts are concentrated on automotive seating, other applications are being studied for office and institutional furniture and decorated rigid parts. Many different processes have been developed for the 'unit seat' system. 'Unit seat' is the general term to indicate a polyurethane cushion moulded with its cover. The cover can be PVC previously vacuum thermoformed, such as the 'Controform' process,[10] or can be based on low temperature stretchable material preferably fabric vacuum formed at low temperature in the foaming mould, such as the 'Skin-form' process.[11] In both processes the mould cavity is connected with a vacuum chamber.

An automatic approach to form-and-foam processing is 'Skin-form'. Moulds are automatically moved from station to station by a multi-pallet conveyor. A precut cover of composite textile or composite vinyl is placed over the mould cavity and held in place by a clamping frame fitted with spring-supported ball bearings on the underside to permit controlled cover

FIG. 15. Seats produced by 'Skin-form' process.

movement. The cover is positioned and vacuum applied. The vacuum is applied separately to the sides as well as to the bottom of each mould and it is held until the end of the foam curing. Some seats made by 'Skin-form' process are shown in Fig. 15.

REACTION INJECTION MOULDING EQUIPMENT

Low Pressure to High Pressure Techniques

The development of reaction injection moulding (RIM) and its increasing commercial application of large parts for the automotive industry and also structural foam applications produced significant development of high pressure metering units. Only by using high pressure (HP) machines has RIM technology been commercially developed and then these types of machines have been improved to obtain minimum surge and efficient mixing, thereby enabling self-skinned articles to be produced.

In the development of HP machines, the self-cleaning mixing head has a mixing chamber which is cleaned out by a piston so eliminating any residual material in the head.

The self-cleaning mixing head extended the use of HP machines in other urethane product areas such as cold flexible and semi-rigid foam moulding, rigid foam and shoe soles, where the liquid material is poured in the open mould or through a hole not directly connected with the mixing head. In this last case the mixing head is fitted to the injection point by pneumatic or hydraulic pressure and after filling the hole is automatically closed by a plug.

In the case of pouring in open moulds, a special head with a long extension pipe has been developed. This extension decreases the foaming material pressure, thereby avoiding the spraying of liquid material.

Another application of the high pressure technique is in insulating panels. The panels are filled with rigid foam by complete insertion of a small flat head into the two surfaces of the panel. Figure 16 shows this particular type of head.

Other developments are adaptations of high pressure equipment to facilitate colour changeover and simultaneous production of material with different colours. In some systems colour paste is brought directly to the head; in another system several coloured polyols blends drawn from separate tanks are combined with isocyanate in one or more mixing heads (see Fig. 17).

FIG. 16.   Flat mixing head for pouring rigid foam in insulating panels.

Reaction Injection Moulding Application

Due to automotive industry interest in impact resistant body components, great technical advances in the application of the RIM concept to microcellular elastomeric urethanes have been achieved.

RIM is the result of a marriage between the chemical systems and process equipment and is predominantly used for the production of PU microcellular elastomeric and self-skinned foam. Because much of the elastomeric RIM market is for the production of large moulded parts for the automotive industry, the required short mould cycle time and the complexity of the article shape, often with very thin thicknesses, needed highly reactive chemical systems and also high pressure impingement mixing foam equipment to allow easier mould filling. With the highly reactive chemical systems, articles with sections 100 mm thick can be taken out of the mould in 1 to 3 min. These very reactive foaming materials normally have a cream time of about 4 s and a rise and cure time of about 8 s.

The achieved fast cure time also means that the time for mould cleaning and release agent application had to be speeded up. Research was intensified to obtain self-releasing chemical systems. Today this objective has been nearly obtained and spraying a very small amount of release agent is sufficient to avoid problems in demoulding and cleaning.

Fig. 17. Multicomponent HP equipment for coloured integral.

Fig. 18.   Press mould-holder for RIM process.

Solid urethane material with very fast curing properties, typical of engineering thermoplastics, can be produced by RIM, although special low pressure processing systems have also been developed. The solid material can be demoulded in a time as short as 30 s.[12]

Not all RIM elastomeric microcellular material is used in the automotive market, but RIM opened up another application in the field of shoe-soles with chemicals offering mould cycle times of 1 to 2 min compared with 4 or 5 min required with chemical systems which were poured into the open mould.

The RIM process is based on injecting the material by high pressure machine into a closed mould through a sprue gate connected with the mixing head. The production plants are fundamentally two types: static plant and conveyor based plant. In static plant the mixing head is always connected with the mould, and a self-cleaning mixhead is essential as well as fast-curing material. To achieve an economical production rate, HP metering units with multi-heads have been developed where every head is continuously connected with its mould and the mixhead piston closes the injection point. One metering unit can serve 10 mixing heads. The moulds are placed into mould-holders or between the two platens of a press assuring a constant closing pressure of the same moulds. A typical press for the RIM system is that shown in Fig. 18. The clamping forces can be from 10 to 40 tonnes. The platen dimensions of the press and its clamping force can be varied to produce very large parts as well as small articles. It can have the capability for tilting and swinging through two axes and for presenting the mould in front of the operator in order to facilitate working. It can have variable speed opening which is hydraulically operated, so allowing slow initial opening to reduce risk of part damage and then acceleration to save time. Hydraulic ejectors can be positioned in the lower or in the upper platen or in both.

In the production plant where the moulds are placed on a conveyor, a metering unit with one mixing head is used. The mixing head is automatically moved to meet the mould in the point where the filling sprue is placed and then locked during the injection. At the end of the injection, the self-cleaning mixhead is unclamped and removed, while the injection point is automatically closed by a plug. On the conveyor the mould-holders are positioned so that the mould can be tilted. It is fitted with automatic opening and closing devices.

In the RIM process the mould design is very important; in particular (a) the adjustment between lid and cavity or among the different parts of the mould if a composite mould is used for production, (b) the thickness and shape of the sprue, (c) the distance and angle of the sprue from the moulded article, and (d) the possible presence of an aftermixer before the sprue. The shape of the sprue must control the entry of the material into the mould cavity. It must be constructed so that the velocities of the mixed chemicals are below the level at which turbulent flow occurs, and these velocities are a function of the part thickness which varies with each application and must be studied from mould to mould. The aftermixer is static and may not be necessary, but it is a means of assuring good mixing if some change of physical parameters—viscosity or pressure—occur in the machine. This static mixer is placed just prior to the sprue and must be so designed to effect re-impingement and turbulent mixing of the already initially mixed components.

The moulds must be thermostatically controlled either when placed on the conveyor or in the press where warm water is used. Normally the temperature at the injection point is between 40 and 50 °C. Working with the mould in the press, the optimum temperature for each application can be used. With a tooling approach which cuts cycle time on large moulded parts, the mould is cooled to 20 °C after foaming and then reheated to injection temperature (40 to 50 °C). This system means that about 30 % time saving can be achieved advantageously in static plants, where the cycle time can be heat dependent and not time dependent. Multicavity tooling that permits economical production of small parts has improved RIM's competitive position in comparison with injected thermoplastics. A key requirement is careful design of gates and runners to assure delivery of equal quantities of material to each cavity.

High pressure equipment for the RIM process today gives an accurate, reproducible shot control and mix ratios, low energy consumption, solvent saving due to self-cleaning mixing heads and ease of operation. Metering units with 2, 3 and 4 components are used. A very accurate temperature control keeps the material within a maximum range of $+1$ °C. Better control over cell structure is obtained with air nucleation of the polyol stream: minute air bubbles are introduced by different devices to decrease the polyol viscosity and bring it nearer to that of the isocyanate component and to act as a nucleating blowing point so that a homogeneous cell structure is obtained when microcellular material is produced.

Key features of reproducible shot control of these high pressure machines include by-pass recirculation with low pressure recirculation between shot and shot, precisely adjustable inlet valves, a new design of the piston to empty the mixing head, and shot calibration. The complete inlet valve system, including the nozzle orifice, is directly fitted in the mixhead.

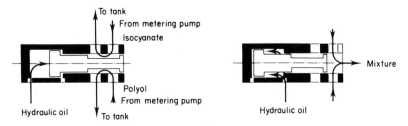

FIG. 19.   High pressure recirculation in the mixing head.

The nozzle orifice is adjustable. A hydraulic piston is incorporated in the mixing head in such a manner as to avoid contamination from residual chemical material. The system of by-pass recirculation presents some differences between the equipment built by different suppliers. A type of mixing head (Fig. 19) has a short by-pass recirculation at high pressure in the mixing head before the shot: the piston has grooves corresponding to the recycling and pouring requirements of the components and the mixhead can be switched from the recycle to the pouring mode by simultaneously uncovering the pour ports through the action of the central hydraulic piston.[13] In one type of mixing head there is a short recirculation under high pressure just before the nozzles (Fig. 20) with a pressure very near to that of pouring. Another system has by-pass recirculation before the nozzles; then the recirculation valves are closed completely or partially in such a way as to obtain a pressure corresponding to that of the impingement when the hydraulic piston uncovers the nozzles for the injection. The delay between the valves closing and the injection is constant and very short.

Recently new equipment to produce filled and fibre-glass reinforced microcellular PU have been studied for the RIM process and introduced in the market. Some equipment has a tank (or two, if the fibre-glass is divided

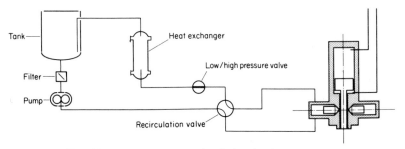

FIG. 20.   Low pressure recirculation in the mixing head.

between the polyol and isocyanate stream) that receives premixed filler and polyol from a pressurised tank. The accumulator is fitted with a hydraulically driven ram transferring the mixture in the required amounts from the accumulator to the mixing head, where it is combined with isocyanate via conventional high-impingement techniques. There is some difference in the pressure of the tank feeding the accumulator: from 3 to 10 bar depending on the type of machine. With a second tank for adding glass fibres in the isocyanate stream (using special grades of milled dry glass fibres) it is possible to increase the glass fibre content. Good results have been obtained with more than 50% of short glass fibres (less than 1 mm), but a level of 10 to 15% of the right size (2 to 3 mm) fibre seems to be sufficient for improved mechanical properties. In other high pressure machines a system of continuous recirculation of material from the tank to the mixhead working at above ambient temperatures (as generally required in ISP, i.e. instant set polymer) seems to give good results; preferably the fibre-glass is divided and added in both the polyol and isocyanate streams.[14]

REFERENCES

1. Tasc Unit, Pye Electric Company Ltd, England.
2. Honeycomb Nozzle, Patent No. 41,484 (1965), Viking Engineering Co. Ltd, England.
3. Draka Square Block Process, Patent No. 96,048 (1958), Draka, Holland.
4. Planiblock Process, Patent No. 423,434 (1973), Spain.
5. Hennecke, Germany—Rectangular Block Device, Patent No. 1,392,859 (1972), England.
6. Maxfoam Process, Patent No. 131,636 (1970), Norway.
7. Varimax Process, Patent Unifoam, Glarus, Switzerland.
8. BUIST, J. M. (1965). Advances in rigid urethane foam process and applications, *J. Cell. Plast.*, **1**(1), 101.
9. DuPONT ELASTOMERS LABORATORY (1976). *J. Cell. Plast.*, **12**(2), Mar./Apr., 74.
10. STOREY BROS. LTD, UK (1975). *Modern Plastics International*, **4**(10), Oct., 25.
11. Skin-form Process—Jose Jover, Spain, and Studio Tecnico Mazzucco–Rubatto, Italy, 4th June, 1970.
12. *Modern Plastics International*, **5**(11), Nov., 1976, 44.
13. PREPELKA, J. and WHARTON, J. L. (1975). *J. Cell. Plast.*, **11**(2), Mar./Apr., 93.
14. *Modern Plastics International*, **7**(12), Dec., 1977, 10.

Chapter 10

HEALTH HAZARDS ENCOUNTERED IN THE INDUSTRIAL APPLICATION OF ISOCYANATES

W. Bunge and F. K. Brochhagen

Bayer AG, Leverkusen, West Germany

SUMMARY

All over the world, isocyanates are being processed in many large industrial companies and also smaller companies so that more than 100 000 people are working with these chemicals. Isocyanates are used to produce foamed and unfoamed polyurethane polymers, lacquers, coatings and adhesives.

The polyurethane industry became established in the early 1950s and from its inception the importance of limiting exposure to isocyanates was recognised. The history of the industry using these reactive chemicals in diverse applications is good. A threshold limit value of 0·02 ppm has been set as the upper limit of concentration for factory exposures which is regarded as safe.

The medical effects of exposure to various concentrations of isocyanates are discussed in some detail and reference is made to the guidelines, laid down in various countries, of the principles to be followed to ensure good industrial hygiene when working with these reactive chemicals.

INTRODUCTION

Polyurethane-based synthetic materials, manufactured with the aid of the polyaddition reaction between isocyanates and compounds containing acid hydrogen which was discovered in 1937, have become extremely important throughout the world. The possibility of producing materials with 'tailor made' properties by varying the starting components and the processing conditions has led to a large number of applications in many different spheres of technology and everyday life. Research workers and application technologists are still finding new ways of varying the design and properties of the end products. New fields of application for polyurethanes are constantly taking the experts by surprise.

The nominal capacity for the two most important products, toluene diisocyanate (TDI) and diphenylmethane diisocyanate (MDI), was 770 000 tonnes and 495 000 tonnes per year respectively (as at end of 1976). These products are used mainly in the manufacture of polyurethane foams. 1,5-naphthalene diisocyanate (NDI) plays an important role in the production

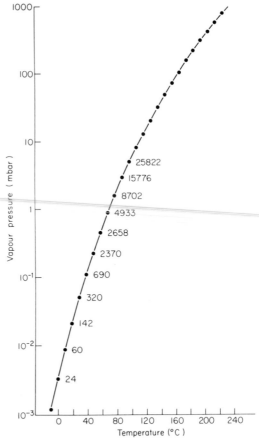

FIG. 1. Vapour pressure curve for toluene diisocyanate. (The figures to the right of the curve are the concentrations of the saturated vapour in mg/m^3.)

of polyurethane elastomers. The aliphatic types hexamethylene diisocyanate (HDI) and isophorone diisocyanate (IPDI) find their chief application in polyurethane coatings, where they are used in a modified form with extremely low proportions of the monomeric starting component. The high reactivity of the isocyanates leads to problems with regard to industrial safety and hygiene. These problems must be viewed in the specific context of particular working conditions, which vary quite considerably, depending on the way in which the isocyanates are handled and processed.

PHYSICAL PROPERTIES OF ISOCYANATES

The properties of TDI and MDI are summarised in Table 1. Figures 1 and 2 show the effect of temperature on the vapour pressures of TDI and MDI and the concentrations of the saturated vapour at some characteristic temperature.

TABLE 1

Property	TDI (80/20)	MDI (polymeric)	MDI (pure)
Physical state at normal temperatures	liquid (thin)	liquid (oily)	—
Viscosity (mPas at 25 °C)	3	200–300	—
Colour	colourless to pale yellow (clear)	dark brown (opaque)	—
Odour	pungent (characteristic)	earthy musty (characteristic)	—
Specific gravity (at 25 °C)	1·22	1·23	—
Boiling point (°C)	250	polymerises about 260 °C with evolution of carbon dioxide	—
Flash point (°C)	127	over 200	—
Fire point (°C)	145	over 200	—
Freezing point (°C)	below 14	below 10	38
Vapour density (air = 1)	6·0	8·5	—
Vapour pressure (mbar at 25 °C)	3×10^{-2}	below 10^{-4}	—
TLV (ppm)	0·02	0·02	—
(mg/m³)	0·14	0·2	—

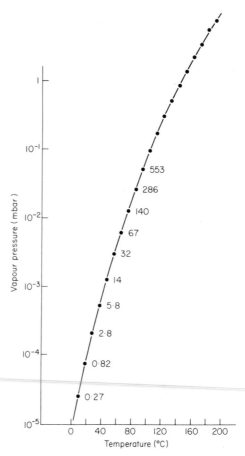

FIG. 2. Vapour pressure curve for diphenylmethane 4,4′-diisocyanate. (The figures to the right of the curve are the concentrations of the saturated vapour in mg/m³.)

TDI is marketed as a mixture of the two isomers 2,4- and 2,6-toluene diisocyanate, the most important type being an 80/20 mixture of 2,4/2,6 TDI, although a 65/35 mixture is also used. The pure 2,4-isomer can be manufactured on an industrial scale and is occasionally used as a component to be combined with other TDI isomers.

In addition to the polymeric MDI oligomer mixture, which is an undistilled mixture of 4,4′-diphenylmethane diisocyanate with isomers and multinuclear derivatives of this compound, pure MDI also plays an important role. Whereas the oligomer mixture is liquid at room temperature, pure MDI is solid (solidification point 38–39 °C) and appears on the market in flake or molten and solidified form.

NDI is a solid product with a solidification point of 127 °C, which is sold in powder form. HDI and IPDI, which are liquid at room temperature, are only occasionally supplied as such to the market, usually being sold in modified form.

HANDLING ISOCYANATES

The aim of this chapter is to outline some of the typical conditions under which isocyanates are handled by the processing industry and to describe medical hazards and their control.

Because of their exceptional versatility and wide variety of forms, polyurethanes are manufactured in numerous large industrial concerns, but to an increasing extent in small factories as well. As a result, there are today more than 100 000 people throughout the world involved in the processing of isocyanates, whether it be in the production of foamed and non-foamed polyurethane (PUR) plastics, in the manufacture of PUR-based paints, coatings and adhesives or any of the several other specific applications of isocyanates.

In the course of their work there is a potential for exposure to isocyanate vapours by inhalation but people should not be exposed to these risks if adequate precautions are taken. The vapour pressure of the isocyanate, the temperature resulting from the exothermic reaction of the isocyanate with hydroxy groups, and the elevated processing temperature of the components which is frequently necessary all affect the concentration of the vapour.

Exposure to aerosols containing isocyanate is conceivable during spraying of reaction mixtures. Apart from this, contact with the skin can only result from improper handling and oral ingestion should never happen.

Plants which manufacture isocyanates present a special situation in view of the starting materials used (aromatic amines and phosgene), and to discuss in detail this aspect as well would be beyond the scope of this chapter.

Manufacture of Polyurethane Blocks, Sheets and Films by Continuous Processes

Blocks made of flexible PUR foam, mostly based on TDI, blocks and sheets of a defined thickness made of rigid PUR foam, usually based on MDI, and PUR coatings for textile webs or plastics film are manufactured in large quantities by continuous methods. The employees can easily be protected from the isocyanate vapours liberated during the manufacture of flexible and rigid foam by installing appropriately designed, permanent exhaust ventilation units. The amount of isocyanate handled per unit of time must be taken into account. In the case of flexible foam this may be up to 150 kg/min. In the production of rigid foam sheets on continuous laminators, the amounts are usually only about one-tenth of this.

Some of the systems used in the continuous coating of webs with homogeneous or cellular polyurethanes include solvents. As the modified isocyanates used for these systems contain the monomeric base products in only extremely low concentrations—usually less than 1 %—the measures to be taken in the processing plants are to be geared especially to the type and amount of solvent used.

Discontinuous Manufacture of PUR Mouldings

Large quantities of shaped components made of PUR foam are manufactured in moulds, for example:

—flexible foam mouldings without facing for upholstery;
—moulded parts manufactured from rigid or semi-flexible foam by the so-called RIM process with a continuous PUR outer skin (integral skin foam) produced in the same process. These mouldings are used mainly in the furniture, automotive, building, sports goods and shoe industries;
—shaped articles made of PUR foam with facings based on inorganic materials or made of plastics, e.g. refrigerators, building components and automobile parts.

The manufacturing techniques depend on:

—the type, size and shape of the item to be produced;
—the reactivity of the raw materials used and the mould residence time, which is influenced to a large extent by these;
—the preheating of moulds and the preheating or afterheating of shaped articles which is necessary in many cases;
—the number of components required in each case.

The devices installed for the exhaustion of liberated isocyanate vapours have to be designed to suit the individual circumstances. In most cases there will be several exhaust points and it must be borne in mind that the concentration of isocyanate in the workroom air fluctuates in accordance with the production cycles.

Manufacture of Polyurethane at the Place of Use

If PUR rigid foams and coatings are not factory-produced, exhaust ventilation may be difficult to install and employees will then require to be protected by wearing an air-fed mask as well as other appropriate protective clothing. Special attention must also be paid in situations of non-factory production to aerosols containing isocyanate which form during the manufacture of PUR foams by the spraying method.

It should be noted that filter masks are unsuitable as they soon become blocked up with aerosol particles.

Isocyanate-based Paints and Adhesives

The problem of exposure to isocyanate does not really arise in the processing of paints and adhesives by the classic brushing method, as either the products already mentioned with an extremely low monomer content are used or, especially in the case of adhesives, monomers such as triphenylmethane triisocyanate, which have an extremely low vapour pressure, are employed. In processing by the spraying method, the protective measures to be taken are the same as those generally necessary for this application technique to prevent inhalation of aerosol, namely exhaust ventilation and/or the provision of respirators. The type and amount of solvent used must always be taken into account.

ANALYTICAL METHODS

The polyurethane industry became established in the early 1950s and from its inception the importance of limiting exposure to isocyanates was recognised. It is to the credit of the industry using these reactive chemicals that by the late 1950s a Threshold Limit Value (TLV) of 0·1 ppm was set for TDI concentrations. An extended examination of the health hazards connected with isocyanate exposure led to a reduction of the TLV to 0·02 ppm and, in 1971 after further study, medical experts confirmed that this level should be regarded as the upper limit of concentration for factory exposures.

Various analytical methods have been developed to detect such low concentrations in the air.[1-5] One method uses atmospheric samples collected by passing air through absorber solutions. The absorbed isocyanates are then converted chemically into colour complexes, which are compared with a standard solution or measured photometrically. The colour intensity is converted to give the isocyanate concentration. A disadvantage of this method is that the sampling takes a considerable amount of time to collect a sufficient quantity of isocyanate to be analysed precisely.

An alternative method is that in which average air concentrations of TDI and MDI over a period of 5 min can be determined by means of a continuous method based on a dry colour reaction.[6] The disadvantage of both methods, apart from the time taken for sampling (about 20 min), is that the apparatus required does not permit the measurement of the air at the operator's breathing level.

These analytical methods can be used very successfully in continuous PUR manufacturing processes, where the air concentrations, taken over longer periods of time, barely fluctuate. In discontinuous processes, however, which are being used to an increasing extent, the isocyanate concentrations naturally fluctuate in accordance with the cycle of the manufacturing procedure. New, considerably more sensitive measuring techniques, with which the sampling can be completed in less than a minute, are of interest because some industrial hygienists claim that the TLV should not be exceeded even in the short-term.

The first successes in this direction were achieved with a modified version of a thin layer chromatographic method,[7] which had already been used in practice for some time and which permits the determination of concentrations of as little as 0·005 ppm at a sampling time of less than a minute. This method also makes it possible to determine several isocyanates together. The variation described in ref. 8 is based on the same chemical principle, but uses a different separation technique (high-pressure liquid chromatography).

In the meantime, intensive research is being carried out into the possibility of developing other methods, some based on purely physical measurements. Apart from reducing the time required for sampling, the aim of these studies is to find suitable evaluation techniques which can be applied in a short time and without a great deal of expenditure and work at the place of measurement itself.

With the aid of these analytical processes it is possible to recognise when there is a risk of isocyanate affecting those working with it, and to estimate the extent of this risk, on the basis of the concentrations measured at the place of work itself. In addition, they are extremely useful in testing the level and constancy of different concentrations when ascertaining the acute and chronic toxicity on inhalation in experiments with animals.

MECHANISM OF THE ACTION OF ISOCYANATES ON BIOLOGICAL STRUCTURES

Isocyanates—depending on their vapour pressure and their method of use (in certain processes they occur as aerosols)—are strong irritants acting mainly on the mucous membranes of the upper and middle respiratory tract and on the cornea and conjunctiva of the eye. This action is due to the ability of the isocyanate groups to react with biochemical structures via their reactive hydrogen atoms thus leading—in relation to concentration and time—to irritation, or sometimes to lesions in the form of cell and tissue necrosis. In extreme cases, a kind of 'hardening' of the affected tissue can be observed.

Action on the Skin

If the liquid compound acts on the skin, lesions occur only rarely following continuous or repeated contact, since, after the reaction of the isocyanate groups with the outermost layers of the skin, a deeper penetration into the living strata of the multi-layered dermal epithelium is largely prevented. Skin diseases resulting from the action of isocyanate pose no occupation-related toxicological problems apart from the rare cases of sensitisation of the skin. The cases of skin sensitisation that have been observed are mostly related to the auxiliaries used, e.g. tertiary aliphatic amines.

Even large accidental spillages of TDI on the skin of a worker, provided it is washed off immediately, has not caused a primary irritation dermatitis.

Oral Ingestion

The acute oral LD_{50} (rat) of TDI is 5800 mg/kg, the acute oral LD_{50} (rat) of crude MDI is 15 g/kg.[9,10] Swallowing of industrial chemicals at work is, however, a rare occurrence. Should it occur—even deliberately—first aid and treatment should be by stomach washout and further medical surveillance as judged clinically.

Inhalation

The main risk to the health of men working with isocyanates is the inhalation in the form of vapour or aerosol. The effects of high and low concentrations are discussed below.

High Concentration—Acute and Subchronic Effects

Acute effects of exposure to high concentrations are irritation of the eyes and respiratory tract. Chemical bronchitis and bronchopneumonia may occur but only rarely have any cases of lung oedema been reported. With modern plant and working methods, such severe acute effects are only likely following an accidental exposure or when all recommended or prescribed protective measures have been disregarded. In severe cases, complete healing of the acute condition by regenerative processes is possible,[11] especially if medical treatment begins early.

Especially in the 1950s and 1960s[12-16] a number of acute cases of disease were described in the literature which were related to exposure to high concentrations of isocyanates. These were connected with either accidents or careless handling of the substances. Lack of knowledge about the dangers of isocyanates at such high concentrations often played a major role in causation because work was carried out without adequate personal protection. These diseases were mostly induced by the TDI used in the manufacture of polyurethane soft foam. A few cases also became known in connection with work with MDI at elevated temperatures. Further, a few cases of intoxication following the inhalation of NDI aerosol were described.[16a]

In the use of the components for polyurethane surface coatings, symptoms occurred if the isocyanate component had a high monomeric content (in part 10% or more).[13,15,16] When the proportion of monomers had been lowered to 0·7%, this hazard could be eliminated from the painting of the coatings. However, the increased use of spraying again gave rise to medical problems for the workers. With this form of application all the components of the coating are released in form of an aerosol into the ambient air and can be inhaled by the worker unless appropriate protective measures are taken. It should be noted that this is not primarily an isocyanate problem but arises with the spraying of coating systems generally. Good industrial hygiene at the work place is vital for the protection of health.[17]

Further knowledge of the local inhalation toxic action of isocyanates (in animal experiments) in relation to atmospheric concentrations has been reported in the literature.[9,18-20]

Action of Low Concentrations—Subchronic Effects

Besides acute intoxication, the pattern of disease following the repeated exposure to low concentrations is of great importance.

Exposure to low concentrations generally causes only subjective symptoms such as burning eyes, dry throat and coughing irritation, but normally the work is seldom interrupted. It is known, however, that those frequently exposed to the inhalation of isocyanates at lower concentrations after several years are more often affected by diseases of the upper respiratory tract (e.g. chronic bronchitis or allergic asthma). The action of isocyanates over a period of several years at these levels of concentration, can lead to progressive impairment of pulmonary function with considerable shortness of breath and stress on the heart.[13,15,16,22,23,24,28] The question of pulmonary damage with subchronic exposure to low isocyanate concentrations has been studied in animal experiments.[19,21,25,26]

It follows that there is a need for the control of atmospheric isocyanate concentrations at the work place to levels which on long-term exposure are not detrimental to the health of the workers.

Increased Sensitivity and Allergy

Besides these diseases of the respiratory organs due to the specific irritant action of the isocyanate, typical bronchial asthma has been observed with some workers. After a first inhalation of isocyanate an asthmatic reaction can be elicited by a subsequent inhalation of much lower concentration. Whether in every case this is actually an allergic process in the form of an immunobiological reaction has not yet been clarified in spite of extensive investigations. Some authors[27-30] found some specific antibodies in a proportion of the workers exposed to isocyanates after both short or long periods. A corresponding clinical manifestation of an allergic reaction was not observed in every case. In apparently 'sensitised' subjects, no relation was found between antibody formation and a corresponding blood and/or sputum eosinophilia.

Other authors[31] were unable to demonstrate specific circulating IgE antibodies against TDI in the blood; the tests showed no differences between 'sensitised' and 'non-sensitised' subjects. However, still under discussion is the possibility of intra-pulmonary localised IgE antibodies which pass into the blood only at such minor concentrations that they escaped detection by the available methods. The reaction between these antibodies and TDI via release of pharmacological mediators could be responsible for the bronchoconstrictor action of this isocyanate. Further investigations in this direction have been planned. The investigators also discuss the possibility that TDI disturbs the function of β-adrenergic receptors—maybe by blockade of β-adrenergic receptors—thus inducing a preponderance of bronchoconstrictor mediators. A fact suggestive of an allergic process from the clinical aspect is that, besides an immediate reaction, some affected subjects developed asthmatic symptoms only hours after the contact with isocyanate—a delayed reaction.

Pepys[32] found in inhalational provocation tests on sensitised subjects with TDI, MDI and HDI that in some cases a cross allergy seemed to exist between the various isocyanates, whereas others showed a reaction only to one or two isocyanates in the same tests. There is no doubt that not all asthmatic reactions after contact with isocyanate are allergic. Especially in a predamaged system—whatever the cause of the damage may have been— symptoms can occur with the inhalation of any irritant even at low concentrations. Recurrent bacterial infections of the bronchial system are also more frequent in these people.

The possible sensitisation of some employees led to the practical decision to arrange alternative work which would not expose allergic and/or bronchitic men to isocyanates. Subjects once sensitised to isocyanates should not have any further contact with these substances.

A question of particular importance in industrial medicine is whether and by what means it can be determined in advance of starting work if a risk of sensitisation to isocyanates exists. The preliminary reports of a US study[31] appear to suggest some possibilities. In a provocation test with acetyl-β-methylcholine (mecholyl) seven out of ten TDI 'sensitised' subjects showed a positive reaction, i.e. a 20% or greater decrease of forced expiratory volume (FEV) whereas of 'non-sensitised' subjects only one out of ten showed a comparable response. This may indicate a 'pre-employment screening' method which in future may enable workers at risk to be recognised in advance of starting work and thus be protected from a possible sensitisation against isocyanates.[31]

Percutaneous Absorption of Isocyanates
It is unlikely that isocyanates will be absorbed through healthy intact skin and so systemic toxic effects will not occur by this route. This opinion was confirmed both by the results of animal experiments and the negative outcome of examinations (clinical examination, haematological, clinical chemical and enzyme-diagnostic laboratory tests) of a large group of subjects.[33] Toxic effects in workers described in the literature,[36] i.e. toxic fatty hypertrophy of the liver are not due to the isocyanate but to the solvents or other chemicals used in the particular manufacturing process.

LONG-TERM EXPOSURE

Threshold Limit Concentrations at the Work Place

Knowledge of the health hazards arising from isocyanate vapours or aerosols has led in many countries to the specifying of threshold limit concentrations at the place of work for the protection of the workers' health. In addition to the already mentioned observations on humans, the results of animal experiments have enabled the assessment of these limits (TLV—Threshold Limit Values—in the USA and the United Kingdom, and MAK—Maximale Arbeitsplatzkonzentration—in the Federal Republic of Germany[34,35]) to be set. In Germany the MAK for isocyanates is considered as the upper limit and not as the mean concentration.

As a result of animal inhalation experiments in different species (mice, rats and guinea pigs) together with experience and examinations at the work place over many years, the TLV for diisocyanates has been set at 0·02 ppm. It is widely accepted by occupational physicians who are experienced in the case of isocyanate workers that atmospheric concentrations below the TLV can result in respiratory sensitisation of a small number of very highly susceptible individuals. It is also the opinion of these physicians that the current TLV is adequate to safeguard the health of the great majority of people.

Long-Term Studies at and below Isocyanate TLV Value Concentrations

It is only fair to record varying experience in recent publications concerning the question of whether or not isocyanates at concentrations below the TLV value of 0·02 ppm cause impairment of pulmonary function. The results of the investigations carried out in recent years by the working group around Peters[37–40] led the authors to believe that measurable losses of pulmonary function (drop in FEV) are present in workers in polyurethane foam production over a period of many years.

Other authors (Adams,[41] Ehrlicher and Brochhagen,[33] Butcher et al.[31]) studying much larger groups of people, with many more measurements at the work place, did not obtain similar results. They found no statistically significant alteration of pulmonary function after exposure for many years. Adams[41] showed that 76 men worked under good work place conditions with TDI for over 9 years without any impairment. One group of workers who had to change their job because of clinical symptoms, did, however, show that long-term loss of pulmonary function can occur.

Ehrlicher[33] states in one study that in a group of 341 individuals no significant differences were found in the parameters of lung function between workers exposed to diisocyanates for 10–15 years and those exposed for 1–3 years. The values matched those of a comparable group not exposed to isocyanates. An average level of 0·0197 ppm isocyanate (TDI, MDI, NDI) was calculated for the work places on the basis of 159 air analyses. Many samples were about or below the sensitivity of the method, i.e. 0·0001 ppm. However, peak exposure rates of 0·4 ppm, 0·6 ppm and in one case even 1·3 ppm were also measured.

A further long-term study by Butcher *et al.*[42] shows that, apart from a certain group of workers in whom sensitisation to TDI was found, no loss of pulmonary function was observed in TDI exposed subjects, compared with a non-exposed control group.

Work Place Industrial Hygiene
Worldwide experience at very different types of work place involving the use of isocyanates has shown that isocyanates can be handled safely if the TLV value of 0·02 ppm is not exceeded. The TLV for TDI of 0·005 ppm proposed by the US Department of Health, Education and Welfare as time-weighted average and a 20 min limit for a maximum allowable concentration of 0·02 ppm is essentially consistent with the level stated above and from practical aspects does not represent an additional safety factor. Acute exposure to atmospheric concentrations of monomeric isocyanates below the TLV value will not lead to long-term lung damage due to irritant effects. When an employee has developed respiratory sensitisation to isocyanates, it is no longer safe to expose him to even the TLV level. The affected employee must avoid any further contact with isocyanates. However, how-far the TLV value provides reliable protection for persons who suffer from constitutional or acquired hyper-activity of the bronchial system (e.g. due to recurrent bacterial bronchitis or after inhalation of relatively high concentrations of irritant gases, aerosols or dusts) can only be decided for each individual case. Principally, in such cases, contact with irritant substances or dusts—isocyanates included—should be avoided.

Pre-employment and subsequent medical assessments have been described in the US by the NIOSH[43] and in the Federal Republic of Germany[44] by the union of the employer's legal liability insurance association and in the UK by the BRMA.[45]

In factories, the TLV value must be met by maintaining equipment at high standards and by technical safety and protective measures. The worker using isocyanates must be instructed to work cleanly and be safety conscious. The workrooms must be well aired and ventilated. In addition, at particular points where concentrations could be high, local exhaustion must be installed in order to remove high isocyanate concentrations. Only when optimum use of these measures is not sufficient to meet the TLV value are additional measures of personal protection necessary. An example is spraying at high isocyanate concentrations when aerosols will form. Depending on the aerosol concentration and the duration of the work, air-fed hood masks must then be used.

For some isocyanates the continuous estimation of the room air concentration is possible together with local estimations for limited periods. These short-duration measurements, in particular, enable faults in technical safety or leaks to be demonstrated and corrected.

CONCLUSION

The growing importance of isocyanates, especially TDI and MDI, in industry means that more and more people work with these highly reactive products at their place of work. The isocyanate manufacturers consider it important to carry out further research into the biological activity of this group of substances and to inform the users of the possible health risks if they are not handled properly. In particular the International Isocyanate Institute, to which 25 isocyanate manufacturers in the Americas, Europe and Japan belong, has set itself this task.

REFERENCES

1. MARCALI, K. (1957). *Analytical Chemistry*, **29**, 559.
2. MEDDLE, D. W., RADFORD, D. W. and WOOD, R. (1969). *Analyst*, **94**, 1369.
3. MEDDLE, D. W. and WOOD, R. (1970). *Analyst*, **95**, 402.
4. PILZ, W. (1965). *Mikrochimica Acta*, 687–98.
5. PILZ, W. (1970). *Mikrochimica Acta*, 504–11.
6. TDI-Detector, Model 7000, Universal Environmental Instruments (UK) Ltd, Poole, Dorset BH17 7RZ, England.
7. KELLER, J., DUNLOP, K. L. and SANDRIDGE, R. L. (1974). *Analytical Chemistry*, **46**, 1845.
8. DUNLOP, K. L., SANDRIDGE, R. L. and KELLER, J. (1976). *Analytical Chemistry*, **48**, 497.
9. KIMMERLE, G. (1970). Unpublished animal experiments, Toxicological Institute of Bayer AG, Wuppertal.
10. KIMMERLE, G. (1971). Unpublished animal experiments, Toxicological Institute of Bayer AG, Wuppertal.
11. FRIEBEL, H. and LÜCHTRATH, H. (1955). *Arch. Exp. Path. u. Pharmakol.*, **227**, 93–110.

12. FUCHS, S. and VALADE, P. (1951). *Arch. Mal. Prof.*, **12**, 191.
13. REINL, W. (1953). *Zbl. Arbeitsmed.*, **3**, 103–7.
14. CHLUD, K. and BAZANT, F. (1963). *Wiener Med. Wschr.*, **113**, 584–6.
15. SCHURMANN, D. (1955). *Dtsch. Med. Wschr.*, **80**, 1661–3.
16. SWENSSON, A., HOLMQUIST, C. E. and LUNDGREN, K.-D. (1955). *Brit. J. Industr. Med.*, **12**, 50.
16a. REINL, W. F. (1974). Schnellbächer, *Zbl Arbeits Med.*, **4**, 106.
17. BUNGE, W., EHRLICHER, H. and KIMMERLE, G. (1976). Arbeitsmedizinische Aspekte der Verarbeitung von Lacksystemen im Spritzverfahren, *Schriftenreihe Zbl. Arbeitsmedizin, Arbeitsschutz und Prophylaxe*, **4**, Verlag fur Medizin Dr E. Fischer, Heidelberg.
18. ZAPP, J. A. (1957). *Arch. Industr. Hlth.*, **15**, 324.
19. HENSCHLER, D., ASSMANN, W. and MEYER, K. O. (1962). *Arch. Toxikol.*, **19**, 364.
20. DUNCAN, B., SCHEEL, L. D., FAIRSCHILD, E. J., KILLENS, R. and GRAHAM, S. (1962). *Amer. Industr. Hyg. Ass. J.*, **23**, 447.
21. BUNGE, W. (1976). *Inaug. Diss.*, Wurzburg.
22. SEIDEL, H. and POHLE, H. (1960). *Tuberk.-Arzt.*, **14**, 675–86.
23. MUNN, A. (1965). *Ann. Occup. Hyg.*, **8**, 163.
24. HAMA, G. M. (1957). *Arch. Industr. Hlth.*, **16**, 223.
25. NIEWENHIUS, R., SCHEEL, L., STEMMER, K. and KILLENS, R. (1965). *Amer. Industr. Hyg. Ass. J.*, **26**, 143.
26. KOHLER, R. (1970). *Med. Inaug. Diss.*, Wurzburg.
27. SCHEEL, L. D. (1972). Immunologic changes in men following isocyanate exposure. Read before the Skytop Conference on Respiratory Disease in Industry, Skytop, Pa (see Ref. 43).
28. KONZEN, R. B., CRAFT, B. F., SCHEEL, L. D., and GORSKI, C. H. (1966). *Amer. Industr. Hyg. Ass. J.*, **24**, 121–7.
29. TAYLOR, G. (1970). *Proc. Roy. Soc. Med.*, **63**, 379.
30. CIRLA, A. M., ZEDDA, S. and NAVA, C. (1972). Medichem. First International Symposium, Ludwigshafen.
31. BUTCHER, B. T., SALVAGGIO, J. E., O'NEIL, C. E., WEILL, H. and GARG, O. (1977). *J. Allergy Clin. Immunol.*, **57**, 223.
32. PEPYS, J. (1976–77). Unpublished experiments, Brompton Hospital, London.
33. EHRLICHER, H. and BROCHHAGEN, F. K. (1976). Health problems in the industrial use of isocyanates. Plastics & Rubber Institute Conference *Urethane and the environment*, London, 21–22 Sept.
34. *Threshold Limit Values for Chemical Substances in Workroom Air*, 1976, American Conference of Governmental Industrial Hygienists (ACGIH).
35. *Maximale Arbeitsplatzkonzentrationen*, 1976, Deutsche Forschungsgemeinschaft, Harald Boldt Verlag, 5407 Boppard/Rhein.
36. *Toxikologisch-arbeitsmedizinische Begründungen von MAK-Werten*, 1975, Verlag Chemie, 6940 Weinheim/Bergstrabe.
37. PETERS, J. M., MEAD, J. and VAN GANSE, W. F. (1969). *Amer. Rev. Resp. Dis.*, **99**, 617.
38. PETERS, J. M., MURPHY, R. L. M. and FERRIS, B. G. (1969). *Brit. J. Industr. Med.*, **26**, 115.
39. PETERS, J. M. (1970). *Proc. Roy. Soc. Med.*, **63**, 372.

40. WEGMANN, D. H., PAGNOTTO, D. L., FINE, L. J. and PETERS, J. M. (1974). *J. Occ. Med.*, **16**, 258.
41. ADAMS, W. G. F. (1975). *Brit. J. Industr. Med.*, **32**, 72.
42. BUTCHER, B. T., JONES, R. N., O'NEIL, C., GLINDMEYER, H. W., DIEM, J. E., DHARMARAJAN, V., WEILL, H. and SALVAGGIO, J. E. (1977). *Amer. Rev. Resp. Dis.*, **116**, 411–21.
43. US Department of Health, Education and Welfare. Criteria for a recommended standard . . . . Occupational exposure to toluene diisocyanate. US Government Printing Office: 1973-759-580/1490, Washington DC.
44. Berufsgenossenschaftliche Grundsatze für arbeitsmedizinische Vorsorge- untersuchungen: G 27 Gefährdung durch Isocyanate, Hauptverband der gewerblichan Berufsgenossenschaften eV, Bonn, A. W. Gentner Verlag, Stuttgart.
45. BRMA Code of Practice 1977.

APPENDIX: MEMBER COMPANIES OF THE
INTERNATIONAL ISOCYANATE INSTITUTE, INC.

Western Europe
Bayer AG, Leverkusen, West Germany.
DuPont De Nemours International S.A., Geneva, Switzerland.
Eurane, Paris, France.
Imperial Chemical Industries Ltd, Manchester, UK.
Montedison S.p.A., Milan, Italy.
Prod. Chim. Ugine Kuhlmann, Levallois-Perret Cedex, France.
Rhône Poulenc Industries, Paris, France.
Shell International Chemical Company Ltd, London, UK.

North America
Allied Chemical Corporation, Morristown, New Jersey.
Dow Chemical Company, Midland, Michigan.
E.I. DuPont De Nemours & Company, Wilmington, Delaware.
Mobay Chemical Corporation, Pittsburgh, Pennsylvania.
Olin Corporation, Stamford, Connecticut.
Rubicon Chemicals, Inc., Geismar, Louisiana.
Union Carbide Corporation, New York.
Upjohn Company, Kalamazoo, Michigan.

South America
Isocianatos do Brazil, Sâo Paulo.
　Petroquimica Rio Tercero.

Far East
Mitsubishi Chemical Industries Ltd, Tokyo, Japan.
Mitsui Toatsu Chemicals, Inc., Tokyo, Japan.
Nippon Polyurethane Industry Company Ltd, Tokyo, Japan.
Nippon Soda Company Ltd, Tokyo, Japan.
Sumitomo Bayer Urethane Company Ltd, Amagasaki, Hyogo Pref.,
    Japan.
Takeda Chemical Industries Ltd, Osaka, Japan.

INDEX